FLOORS AND THEIR MAINTENANCE

J.K.P. EDWARDS
T.D., M.A., M.I.I.M., M.B.I.M., F.B.I.C.Sc.

Technical Director
Russell Kirby Limited

CHICHESTER
BEAVER PRESS
CRESTA PUBLISHING

PUBLISHED BY:
THE CRESTA PUBLISHING COMPANY,
14 Beechfield Road,
Liverpool, L18 3EH.

PRINTED BY:
BEAVER PRESS,
Unit 7, Holmdale,
Fontwell Avenue,
Eastergate, Chichester,
West Sussex, PO20 6RY.

First published in 1972
Second impression 1974
Third impression 1981
Fourth impression 1982
Fifth impression 1983
Sixth impression 1984
Seventh impression 1986
Eighth impression 1987

© J K P Edwards, 1972

All rights reserved. No part of this publication may be reproduced or transmitted in any form or by any means, including photocopying and recording, without the written permission of the copyright holder, application for which should be addressed to the publisher. Such written permission must also be obtained before any part of this publication is stored in a retrieval system of any nature.

This book is sold subject to the Standard Conditions of Sale of Net Books and may not be sold in the UK below the net price given by the publishers in their current price list.

ISBN 0 408 70257 5

FOREWORD

In my capacity as Chairman of the Education Committee of the British Institute of Cleaning Science and House Services Manager at Shell Centre, the London Headquarters of the Shell Group of Companies, I have always been conscious of the dearth of authoritative reading matter on the subject of floor maintenance. It is because of this that I welcome this very fine work, which will, I know, rapidly become a standard in its field.

The book deals in some detail with practically every type of floor and contains a wealth of knowledge obviously gained from wide practical experience over many years. It is without doubt the most instructive and useful book yet published on this subject. Full of factual, up-to-date information, it should be in the possession of all those responsible for floor maintenance.

The author is a leading authority on this subject and has written many articles and lectured at both national and international conferences. He is Technical Director of a company manufacturing floor maintenance materials. For his services to education he has been awarded a Fellowship of the British Institute of Cleaning Science.

It is doubtful whether anyone else in the United Kingdom could have written such a competent book on so many different types of flooring. I compliment the author on his initiative and effort in producing a book which will, I am sure, be of immense value to all concerned with floors and their maintenance.

D. L. Williams F.B.I.C.Sc.,
Chairman, Education Committee,
British Institute of Cleaning Science

PREFACE

With the rapid and striking advances which are being made in the wide field of floor maintenance, more and more emphasis is being placed on the need for correctly trained and qualified personnel to carry out these tasks. Well maintained floors are safe and hygienic and contribute in no small way to the total environmental comfort of the whole community.

Clean floors are attractive and correct maintenance procedures protect them from dirt and wear and significantly reduce maintenance costs. Floors represent a very considerable investment and it is in the interest of all concerned that this investment is protected to the best possible extent. It has been estimated that over £600 million are spent annually on cleaning in Great Britain alone. A good proportion of this amount is spent on floor maintenance and it is essential, therefore, that those in responsible positions should be aware of the most modern and up-to-date methods and materials available for this task, so that maximum economies can be effected.

Floor maintenance has developed from a laborious, necessary evil to a modern technology, necessitating scientific knowledge and organisation. Training at all levels is essential if the greatest possible advantages are to be gained from advances in both the manufacture of new types of maintenance material and equipment, and in methods.

In my first book *Floor Maintenance Materials: Their Choice and Uses,* the many types of industrial detergent, floor seal and floor wax are explained in detail. The advantages and disadvantages of each material are discussed, together with the factors that should be considered in their selection. It is emphasised that discrimination in selection of materials should be the policy of all engaged in the maintenance of floors.

This book is a natural sequel. If the correct maintenance materials and procedures are to be applied it is absolutely essential that the nature and characteristics of the type of floor to be treated are understood. It should be recognised that correct maintenance procedures with the wrong materials could permanently damage a floor at worst and prove ineffective at best. Incorrect maintenance procedures with the right materials could also prove harmful to the floor and might cause considerable and unnecessary expense.

PREFACE

In recent years great strides forward have been made in all aspects of floor maintenance. New types of flooring are constantly being developed, many producing their own particular maintenance problems.

The aim of this book is to explain in everyday language how each of the many different floors found in commercial, industrial, public and many other types of building should be maintained. The correct materials to use and avoid and the methods that should be adopted are fully explained. The reasons why certain materials and methods should not be used are also fully considered, so that the reader will be fully aware of the possible consequences of such action. This will also enable faults to be identified so that they can be corrected.

Conversion to the metric system of units is currently in progress, and these are used in the text, generally followed by the old Imperial unit for convenience. Conversion tables also appear in Appendix V.

A Glossary of Technical Terms is also appended to describe, in layman's language, technical terms frequently found in trade journals and literature.

By explaining in some detail the composition, properties and characteristics of each type of floor discussed in the book, it is hoped that a fuller and more complete understanding of the whole subject will be gained, to the benefit of all concerned with floor maintenance.

I would like to place on record my appreciation of the support and constructive advice given to me by Mr L. Harold Russell, J.P., Chairman and Managing Director of Russell Kirby Ltd. I also much appreciate the helpful criticism given by Mr B. G. Taylor, Technical Manager of Russell Kirby Ltd.

My thanks are also due to the Timber Research and Development Association for permission to reproduce diagrams from their booklet *Wood Flooring*.

Some sections have been adapted from articles by myself that have appeared in journals. I am indebted to the Editors of the following journals for permission to reproduce this material: *Caterer and Hotelkeeper; Cleaning and Maintenance; Flooring and Carpet Specifier; Hospital Management; Index to Office Equipment and Supplies; Maintenance and Works Management.*

I am also very grateful to my wife, Lesley, for her patience and care in the preparation and typing of this book.

J. K. P. Edwards

CONTENTS

1	Elements of Floor Maintenance	1
2	Wood Group of Floors	13
3	Stone Group of Floors	38
4	Asphalt Group of Floors	62
5	The Resilient Floors	68
6	Carpet Group of Floor Coverings	94
7	Other Floors	113
Appendix I	Removal of Stains from Carpets	127
Appendix II	Coverage of Floor Seals and Waxes	132
Appendix III	Floor Maintenance Chart	137
Appendix IV	Glossary of Technical Terms	144
Appendix V	Conversion Tables	154
Index		159

1

ELEMENTS OF FLOOR MAINTENANCE

INTRODUCTION

Floors are costly and represent a substantial investment. If floors are neglected they may become damaged, thereby incurring unnecessary expense.

It has been estimated that approximately 80 per cent of dirt entering a building is carried in on the soles of shoes. Maintenance procedures must ensure, therefore, that not only is the dirt effectively removed, but that the floors are also protected from damage.

The cost of maintaining a building over a period of twenty-five years is approximately the same as the original cost of the building. When it is appreciated that the cost of floor maintenance is a considerable proportion of the total maintenance costs, it becomes obvious that correct floor maintenance represents a big investment.

The behaviour of a floor depends as much on the method of maintenance and materials used as on the type of floor. While different types of floor are discussed in subsequent chapters, the aim of this chapter is to outline briefly the main materials used in floor maintenance, namely industrial detergents, floor seals and floor waxes. These materials are discussed in detail in *Floor Maintenance Materials: Their Choice and Uses,* Butterworths (1969).

A knowledge of the different types of material available, together with their properties and characteristics, is essential if the best possible results are to be achieved. Incorrect materials can ruin floors and discrimination in selection is, therefore, essential.

It is, perhaps, convenient at this point to consider briefly the nature of dirt, which has been defined as 'matter in the wrong place'. Dirt consists of widely differing materials. Particulate dirt comprises fine particles of soot, rust, soil and many other items. In general, the smaller the particle size, the harder it sticks and the more difficult it is to remove.

Grease from petrol and diesel oil fumes, with condensation products from chimneys and flues, are very evident in urban and industrial areas. Grease tends to trap and hold particulate dirt and the two types are frequently found together.

ELEMENTS OF FLOOR MAINTENANCE

Organic matter in soilage is widespread and can consist of animal or vegetable debris, or soil laden with bacteria.

INDUSTRIAL DETERGENTS

The subject of detergency is large and complex. With regard to industrial detergents, however, three main classes are readily distinguishable, namely:

Neutral detergents
Alkaline detergents
Caustic detergents.

The class in which a detergent is placed depends upon its pH, or potential of hydrogen.

By using pH, acidity and alkalinity can be expressed in simple numerical terms. The pH scale ranges from 0 to 14. pH 7·0 is neutral and

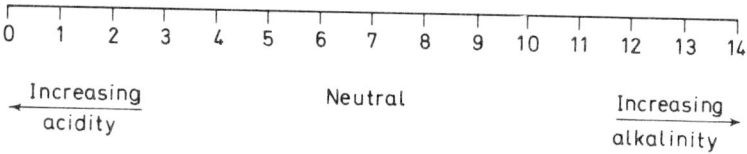

Fig. 1.1. pH scale

is the value given to pure distilled water. Materials with a pH below 7 are acidic, the acidity increasing as the pH decreases. Materials with a pH value above 7 are alkaline, the alkalinity increasing as the pH value increases (*Figure 1.1*).

Although 7 is chemically neutral, in commercial terms materials with a pH value of between approximately 7 and 9 are considered to be neutral. Alkaline materials are normally those within the range from 9 to 12·5 or 13 and caustic materials those with a pH greater than 13. Acids are generally harmful to floors and are not used as routine cleaning materials.

Neutral detergents are used for a wide range of both industrial and domestic applications. They are typified by the liquid detergent used for washing-up, although powder types are also available. The liquid form is generally preferred because of its greater convenience.

Neutral detergents are added to water and when applied to floors they will soften and remove light dirt, soilage and carbon black heel marks, without adversely affecting any floor wax which may be present. Such detergent solutions may be used in conjunction with either mopping equipment or with an electric polishing/scrubbing machine.

One of the main features of neutral detergents is that they are safe for all floors and will not harm surfaces to which they are applied. The use of water, however, on porous floors such as wood should always be minimised and these should first be sealed to ensure that water does not penetrate into the floor itself.

Alkaline detergents are applied where a stronger type of material is required. They are the workhorses of the cleaning industry and can be used in a wide variety of cleaning equipment. Both liquid and powder types are available but, in general, the former are preferred.

Perhaps the main use of alkaline detergents is for the removal of water emulsion floor wax, where they are particularly effective. They will also remove stubborn carbon black heel marks and heavy accumulations of dirt.

Because they are rather strong detergents the floor must be thoroughly rinsed after use to ensure that no detergent remains, otherwise the next application of floor wax may be adversely affected. It is often advisable to add a little neutralising solution, or vinegar, to the rinse water to help neutralise any alkaline detergent remaining. Hands should be thoroughly washed immediately after being in contact with such a detergent. Rubber or PVC gloves should be used if the skin is particularly sensitive.

Caustic detergents are used for applications where a very strong material is required, for example in the cleaning of ovens or of blocked drains. They should not be used on floor coverings because they can have a damaging effect; a solution of caustic soda will cause certain types of green linoleum to turn brown.

It is convenient to include under the heading of industrial detergents two other types of material, namely detergent crystals and solvent-based detergent wax removers.

Detergent crystals, or alkaline degreasers as they are often called, are widely used for industrial applications. They are heavy-duty alkaline materials and are primarily used in conjunction with water for removing oil and grease from both concrete and asphalt floors.

Detergent crystals have very good emulsifying properties and are particularly suitable for cleaning large areas. Because detergent crystal solutions are strongly alkaline, due precautions should be taken to protect the skin and eyes. Rubber or PVC gloves should be worn and if the skin becomes splashed it should be washed in soap and water at once.

Solvent-based detergent wax removers consist of water and solvent, often of the white spirit type, emulsified with powerful wetting agents and other additives. They are designed mainly for the removal of solvent-based waxes and are employed with perfect safety on wood, cork and similar floors. They are also used to remove oil and grease from concrete floors and from equipment and machines.

Because solvent is present they should not be used on floors that may be harmed by solvent, for example asphalt.

FLOOR SEALS

Most floor surfaces require sealing for protective, hygienic and decorative reasons. It prevents the possibility of soilage from becoming embedded in the surface, resulting in difficult removal. Many floors which appear drab and lifeless can be greatly improved by sealing.

A seal can be defined as a semi-permanent material which protects the floor from the entry of dirt, stains and foreign matter. A floor that is sealed is much easier to clean and maintain. Nearly all floors are porous to a certain extent and by filling the open pores and providing a wearing surface, the life of a floor can be extended almost indefinitely. Also, a floor that is sealed and kept clean is far less likely to harbour dirt and germs than one that is neglected.

A seal should penetrate, to some extent, into the surface of the floor, in order to reinforce it and to ensure an even wearing surface. It is also important that the seal should penetrate so that it can 'key' to the floor.

Of a very wide range of floor seals available today, five main types are evident:

    Oleo-resinous seals
    One-pot (ready for use) plastic seals
    Two-pot plastic seals
    Pigmented seals
    Water-based seals.

Oleo-resinous Seals

A typical oleo-resinous seal consists of an oil processed with resin and combined with thinners and driers. Most are penetrating seals. The dried film is tough and flexible. They are very easy to apply and any worn spots can be touched up or renewed, whereas other types of seal require greater care. There is no unpleasant odour, neither has the solvent any ill effects on the hands, provided normal care is taken.

Oleo-resinous seals are rather slower drying than most plastic seals. They are generally hard dry between eight and ten hours, but some require up to 24 hours. They are also rather dark in colour and are not particularly resistant to chemical attack.

Brushes and equipment used to apply the seal can be cleaned with white spirit (turps substitute).

Oleo-resinous seals play an important part in floor maintenance today and are widely used in almost all types of building.

One-pot (ready for use) Plastic Seals

This description is given to those seals which do not contain a drying oil, but dry either by evaporation of solvent or by a chemical reaction.

It is convenient to include the one-pot oil modified and moisture-cured polyurethane seals under this heading.

One-pot polyurethane seals will resist a wide range of industrial and household chemicals, such as solvents and dilute acids and alkalis, although the oil modified types do not have the extreme chemical resistance of a two-pot polyurethane seal. They combine the physical and chemical advantages of a polyurethane with the convenience of being ready for use. The films formed are strong and abrasion resistant. They are also slightly elastic, with the result that they do not become brittle.

One-pot polyurethane seals show very good durability, even under heavy traffic conditions. The seals are easy to apply with either a brush or applicator, although care must be taken to ensure that films are applied thinly and evenly.

Once applied, the seal will dry in approximately three to four hours, the exact time depending upon conditions. In general, however, it will dry quicker than an oleo-resinous seal but not as quickly as a two-pot plastic seal.

Brushes and equipment can be cleaned with white spirit, although with the moisture-cured type of polyurethane a stronger solvent is usually necessary.

Two-pot Plastic Seals

This description is given to those seals which require the blending of two components prior to use. The components are a 'base' containing the main body of the material, and an 'accelerator' or 'hardener' which, when added to the base, initiates a chemical reaction which stops only when the material has fully hardened.

The principle of hardening a resinous film by an accelerator represents an advance over one-pot materials. Better quality seals can be prepared by the two-pot method without any loss of storage stability, but it should be recognised that bases and accelerators must be mixed in the correct proportions in accordance with the manufacturers' instructions. They are frequently supplied in the amounts required for mixing.

Once mixed, the hardening reaction will start immediately. The material must be mixed just prior to use and not allowed to stand for some time in a mixed condition, otherwise it will thicken and may be found unusable when eventually required.

Correct surface preparation is also an essential requirement for success. As the materials dry quickly they do not penetrate to the same extent as oleo-resinous seals and it is, therefore, imperative that the floor is thoroughly clean and dry to ensure maximum adhesion.

Results achieved with a two-pot material amply reward the extra care required.

Two-pot polyurethane seals, in addition to having excellent hardness and abrasion resistance properties, are particularly recommended for their exceptional durability and resistance to chemical attack. They are noted for their low flammability and are probably the best seals available today.

Quick-drying polyurethane seals are hard dry, suitable for overcoating, in approximately two hours. This is a great advantage as it enables a floor to be completely sealed with one primer and two finishing coats and be back in service with the minimum of inconvenience — a very important consideration in all establishments.

The greater durability of polyurethane seals means that the period between recoating operations is considerably extended. This results in less disruption due to clearing rooms and in lower labour costs for actual work on the job. The floor also retains a better appearance over a longer period.

Adequate ventilation is essential to assist drying time and to remove the solvent odour. Special solvents are required for washing brushes and applicators. These should be cleaned immediately after use and not allowed to remain standing in solvent overnight. Care should also be taken to see that accelerator does not come into contact with the skin, as most accelerators are extremely reactive.

The problems of removal at a later date should be considered before sealing is carried out. Two-pot seals are extremely hard and once they have developed their full physical and chemical resistance properties, they are difficult to remove.

The advantages of using a two-pot polyurethane seal are many. Durability and appearance are excellent, chemical resistance is very good and the quick drying time reduces inconvenience to a minimum. Extra care and supervision are, however, required, particularly when mixing the bases and accelerators. A strong odour will also be evident, calling for maximum ventilation. Subsequent removal presents some difficulty due to the excellent adhesion obtained, but if the sealed floor is maintained in a correct manner, touching-up of heavy traffic lanes should only be necessary after a number of years.

Pigmented Seals

A pleasing colour can be given to a floor by the use of pigmented seals. These are widely used on concrete floors, both to improve the appearance and to strengthen the surface layer and prevent it from dusting. As conventional paints break down under action of alkali present in both new and old concrete, special seals are made to combat this effect.

Two main types are available for this purpose. The first is based on synthetic rubber resins and is a one-pot (ready for use) material. This gives a very pleasing appearance and has good durability. It can also be easily renewed when required. It dries in about three hours but it is advisable to leave it to harden overnight before opening the floor to normal traffic.

The second main type is based on polyurethane materials. This seal is superior in all respects to those based on rubber resins. The top-grade pigmented polyurethane seals are of the two-pot variety, so that mixing has to take place before application, but the little extra time required for this operation is well rewarded. Their characteristics and performance are generally the same as the clear two-pot polyurethane seals already discussed.

Water-based Seals

Water-based seals are comparatively new materials, the best consisting of large particle size acrylic polymer resins combined with levelling agents. They are designed for use on porous surfaces which should not normally be sealed with a conventional solvent-based seal. Examples are terrazzo, asphalt, lino, thermoplastic tiles, rubber and PVC (vinyl) asbestos tiles. The seal fills the open pores in the floor and provides a surface which should then be maintained with a water emulsion floor wax.

While not a seal in the strictest sense, it is convenient to include silicate dressings under this heading. The dressings are water-based materials, specifically formulated for use on large areas of concrete. After diluting with water, silicate dressings can be applied through the rose of a watering can, then spread evenly with a soft brush. Of low cost, they react chemically with the materials comprising the concrete, so forming a hard surface and preventing any dusting from taking place. Worn areas can easily be touched up as required.

Surface Preparation

With regard to any sealing operation, it cannot be over-emphasised that correct and thorough surface preparation is essential if the best possible results are to be obtained. Durability of every seal is closely linked with the effectiveness of its adhesion to the floor surface. A seal which will not 'key' properly to the floor will sooner or later become detached and wear off.

The manner in which a floor should be prepared to receive seal depends on a number of factors. As technology in the floor maintenance field progresses, so floor maintenance materials become more complex and correct preparation even more essential. It is false economy to skimp preparatory work. Poor preparation may reduce the life of a seal by months or even years, and in the worst possible circumstances the seal may have to be removed completely and the surface prepared all over again.

It is therefore imperative to give full consideration to the preparation of a floor before any seal is applied, to ensure that the best possible results may be obtained.

FLOOR WAXES

Two main types of floor wax are in use today; solvent-based wax and water-based wax.

Solvent-based wax

Solvent-based wax consists of a suspension of a wax or waxes dispersed in solvents. It is available in both paste and liquid form. Paste wax contains very much more wax than does liquid wax.

Since both paste and liquid wax contain solvent, generally of the white spirit variety, they are flammable and should be kept away from naked lights. The inclusion of white spirit prohibits their use on asphalt, thermoplastic tiles, PVC (vinyl) asbestos tiles and rubber, as solvents can have a detrimental effect on these surfaces. They are, however, widely used on wood, wood composition, cork and linoleum floors.

Paste waxes are intended to be applied thinly and evenly to a clean surface, then buffed. Because of the high wax content, these waxes will not clean a floor and, if applied to a dirty floor, will merely spread the dirt around. They should, therefore, always be applied to a clean surface. A thin coat will give far better results than a thick coat of paste wax. If it is applied thickly, a build-up of thick, glutinous wax can occur which retains dirt, making the floor unsightly and unhygienic.

Once an adequate protective layer of wax has been established, routine maintenance should be carried out using a liquid wax which is a very efficient cleaning agent. This is because the material is free flowing and the solvent component loosens the dirt on the floor as the wax is applied. The loosened dirt is retained in the applicator and a thin film of wax is deposited at the same time. When dry, the film should be buffed to harden the surface and produce an attractive gloss finish.

Liquid wax also resists carbon black heel marks. Any scuff marks which may appear can be readily removed and the gloss restored to its original appearance by buffing.

Results obtained with both liquid and paste wax are very pleasing, providing that the floor is subsequently properly maintained.

Water-based wax

Water-based waxes, perhaps better termed water emulsion floor waxes, have a considerable advantage over solvent-based waxes in that much harder waxes can be incorporated. The wax layer is, therefore, more durable, harder wearing and more resistant to heat than that given by solvent-based waxes.

Early formulations consisted of two main ingredients, a wax and an alkali-soluble resin or shellac. Polymer resins were included at a later date to give increased durability, water resistance and gloss, among other advantages.

The principal ingredients — wax, alkali-soluble resin and polymer — can be blended in almost any proportions to give water emulsion floor waxes with a wide variety of properties.

Because emulsion floor waxes are water based and contain no solvent, they can be used on a much wider variety of floors than solvent-based waxes. They are used successfully on thermoplastic tiles, PVC (vinyl) asbestos, flexible PVC, rubber, asphalt, linoleum, terrazzo and marble, as well as many other lesser known types of floor. They can also be used on sealed wood, wood composition and cork floors, provided that the seal is intact and impervious to water.

Of a very large number of water emulsion floor waxes available, all fall broadly into three main types:

    Fully buffable
    Semi-buffable
    Dry-bright.

Fully Buffable

Fully buffable emulsion floor waxes normally consist mainly of wax. When applied to a floor they dry to a low sheen which can be increased to a high gloss by buffing. Buffing is essential, not only to raise the gloss but also to harden the film and thereby increase durability.

Fully buffable materials generally have good resistance to carbon black heel marks. Perhaps the main advantage of this type of floor wax is the ease of gloss renewal. When the film shows signs of traffic and scuff marks, buffing will quickly restore the appearance and original high gloss. This operation can be repeated as required until insufficient wax remains on the floor and a further coat is necessary.

Semi-buffable

Semi-buffable water emulsion floor waxes normally contain more polymer and less wax than fully buffable materials. They dry to a subdued gloss which can be increased, if required, by buffing. In many instances the initial gloss is considered to be satisfactory and the floor wax is left unbuffed. Buffing, however, hardens the film and prolongs the durability, and initial coats, particularly, should be buffed.

Semi-buffable floor waxes have good resistance to carbon black heel marks and are more scuff resistant than fully buffable materials. Because of the higher proportion of polymer present in semi-buffable materials they are, in general, more durable and have better resistance to water, stains and dirt than fully buffable floor waxes.

Scuff and carbon black heel marks can be readily removed by buffing and the film can be buffed repeatedly to restore the original appearance.

Dry-bright

Dry-bright water emulsion floor waxes generally contain a low proportion of wax and a high proportion of polymer resin. They dry on application to a high initial gloss, without buffing. True dry-brights will show very little, if any, increase in gloss on buffing.

ELEMENTS OF FLOOR MAINTENANCE

These materials are very durable and have excellent resistance to water and stains. They also have extremely good anti-slip properties, because the polymer will not give to the same extent as wax when subjected to foot traffic. For the same reason, resistance to carbon black heel marks is not as good as fully buffable or semi-buffable materials and the black marks are more difficult to remove.

It should be recognised that while the above are the three main types of water emulsion floor wax available today, the majority of products lie between these types and combine the desirable properties of more than one type. For example, many products combine dry-bright with buffable characteristics. In the industrial field, particularly, some degree of buffability is desirable, so that the surface can be easily renewed without the necessity of applying a further coat.

Surface Preparation

Correct and thorough preparation is essential if the best possible results are to be obtained. While in certain circumstances one type of water emulsion floor wax can be used on top of another, in general all previous applications of floor wax should be removed. Removal of all old floor wax is essential if it is intended to change from a solvent-based wax to a water-based wax or vice versa.

Solvent-based wax should be removed by using a solvent-based detergent wax remover. Thorough rinsing is essential after the solvent-based wax has been removed and the floor should be allowed to dry before wax is applied.

A water-based wax should be removed by using an alkaline detergent solution in water. It is always advisable to add a little vinegar to the rinse water to ensure than no alkali remains on the floor. After rinsing, the floor should be allowed to dry before floor wax is applied.

Maintenance of Waxed Floors

Correct maintenance of waxed floors is essential if they are to be kept in a clean, attractive and hygienic condition. It also ensures that maximum durability is obtained from the wax.

Perhaps the greatest benefit to be gained by maintaining floor waxes correctly is a considerable reduction in the amount of time and effort needed to keep them in a clean and attractive condition, with consequent saving in labour and material costs.

It should be recognised that the highest quality materials cannot give optimum results if correct maintenance procedures are not followed.

ELEMENTS OF FLOOR MAINTENANCE

On the other hand, comparatively poor materials can be made to perform well if used correctly.

Selection of the right type of equipment to use for any particular situation is also extremely important. Choice of equipment will depend on many factors, including the availability of trained labour, type of floor and building, and financial and other resources available.

Perhaps the two main types of equipment used today are mopping equipment and floor maintenance machines. It is not within the scope of this book to consider either of these in detail, suffice to comment that the design and range of both types of equipment have made great strides forward in recent years. Many very sophisticated pieces of equipment are now available and development is continuing at a rapid pace.

With regard to methods of maintenance, perhaps the most important advancement that has taken place in recent years is the development of foam cleaning from spray cleaning.

Foam cleaning is rapidly gaining wide acceptance as an efficient and economic method of maintaining a floor treated with an emulsion floor wax. By operating this cleaning system the floor undergoing treatment can be allowed to remain open to traffic, as only a small area is damp at any one time. This system, therefore, can be operated during normal working hours while allowing traffic to pass over the area being cleaned.

Foam cleaning has many advantages over the conventional spray cleaning method. In the foam cleaning system, material is contained in an aerosol attached to a floor polishing machine by means of an aerosol dispenser unit. All equipment is carried on the machine. By operating a lever attached to the machine handle a small amount of foam is dispensed on the floor. Because the material is in foam consistency, the amount dispensed can be controlled and directed accurately. The floor is then cleaned, waxed and buffed to an attractive sheen with a few passes of the machine.

The application of a wax not only keeps the floor looking attractive but also facilitates maintenance and, by protecting the floor itself from wear, appreciably prolongs the life of the floor.

2

WOOD GROUP OF FLOORS

The wood group of floors comprises wood, wood composition, cork and magnesite. Wood composition has very similar characteristics to wood, and it is right and proper that it should be considered under this heading.

Cork is different from wood in many ways and is frequently considered to be in the general category of resilient floors, because it is normally laid in tile form on a sub-floor. While it is recognised that cork can be classed as a resilient floor, it is convenient to include it in the wood group of floors, because the characteristics and maintenance of cork resemble closer those of wood than the other types of resilient floor, for example thermoplastic tiles, PVC (vinyl) asbestos and rubber.

Magnesite, or magnesium oxychloride, is sometimes considered together with other types of floor, such as the stone or hard surface group. This is a perfectly reasonable approach, particularly as in many other countries magnesite contains no wood at all. In the United Kingdom, however, it normally contains a wood filler which, to some extent, influences the materials that can be applied to magnesite and the methods of maintenance. For this reason, therefore, magnesite has been included in the wood group of floors.

WOOD

Wood has been used for floors over a period of many centuries. Even today some floors which are hundreds of years old are still giving excellent service. In earlier times wood was probably used because it was both suitable for the task and in plentiful supply. Its continued use, in competition with many newly developed types of synthetic flooring materials, is ample evidence, if any is needed, of its outstanding natural qualities.

The incomparable beauty of wood is widely recognised as an asset to be prized and worthy of a high standard of maintenance. While neglected and misused wood floors can deteriorate rapidly both in appearance and condition, a well maintained wood floor can remain in an attractive, hygienic and serviceable state almost indefinitely.

WOOD GROUP OF FLOORS

Wood floors are to be found in almost every type of building. They are particularly suitable for halls in public buildings and in schools, gymnasia, dance halls, ballrooms, dining rooms, hospitals, libraries, and many other locations.

Timber is divided into two classes; softwoods and hardwoods. The classification is a botanical one. Softwoods are included in the order Coniferae, or conifers and hardwoods in the order Dicotyledoneae, or broad-leaf trees. Typical softwoods are spruce and Douglas fir. Beech, maple, teak, and oak are typical hardwoods. Although the terms softwood and hardwood are botanical in origin, in practice they apply almost equally well to the relative hardnesses of the two classes.

In general, softwoods are not as resistant to abrasion or impact as hardwoods. They are more suitable for light traffic and are frequently protected with a floor covering, such as linoleum or carpet.

Hardwoods wear longer and more evenly. Many will withstand heavy traffic successfully over a long period of time. Because of their improved resistance to abrasion and their decorative appearance, hardwoods are usually protected with a seal rather than with a floor covering.

While certain hardwoods will resist indentations of almost any sort, they are not completely resistant to stiletto heels. Stiletto heels, particularly worn heels where the centre stud protrudes, probably cause more surface damage to wood floors, and indeed many other types of floor, than almost any other kind of traffic. Damage from stiletto heels can be serious, even where the hardest type of wood is concerned.

Wood consists essentially of two main substances; cellulose comprising about 50 per cent and lignin which comprises about 20 per cent. A growing tree forms growth rings from the centre. The outer rings are called sapwood, as the cells convey sap to the branches and leaves. It is usually lighter in colour than the centre, or heartwood. The latter is generally more dense than sapwood and less permeable to moisture.

Types of Wood Floor

In general, wood floors can be divided into four main types:

    Strip and board
    Wood block
    Parquet and mosaic
    End-grain paving.

A general knowledge of the construction of each type of floor is important when considering methods of maintenance. This is particularly so when sanding the floor is contemplated, as this removes about 2 mm (1/16 in) of floor. A thin parquet or mosaic floor, for example, could be ruined by sanding right through the wood to the sub-floor, if precautions are not taken.

Each type of floor will be considered briefly.

Strip and Board

Hardwood strip and board floors are produced mainly tongued and grooved, in widths of up to 100 mm (4 in) for strip and over 100 mm (4 in) for board. The lengths of wood are generally end matched, that is the ends are tongued and grooved and fixing is normally to wooden bearers or joists.

Softwood strip and board flooring, however, is not normally end matched. Lengths of strip and board are usually sawn to form heading joints.

Strip flooring is generally secured to the sub-floor, wooden bearers or joists, by an operation called secret nailing, in which nails are driven at an angle of about $50°$ through the board just above the tongue. If

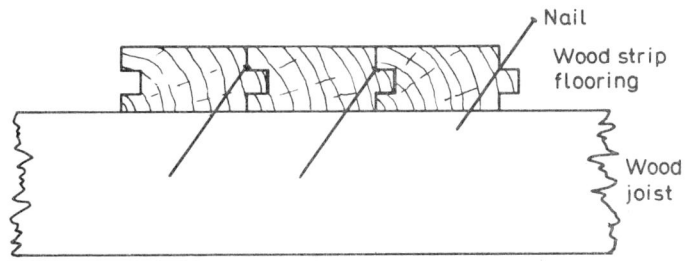

Fig. 2.1. The position of nails in a wood strip floor secured to a joist by secret nailing

sanding a worn strip floor is contemplated it is important to be aware that these nails, though not visible, are, in fact, present (*Figure 2.1*).

Hardwood board flooring is mostly secured by the method used for strip flooring, with additional nails through the face of the wood. Board widths in excess of 175 mm (7 in) are often screwed instead of nailed.

Softwood board flooring is normally face nailed and is intended to be covered by some form of floor covering, such as linoleum or carpet.

Wood Block Floors

Wood blocks are made in a variety of sizes, usually in widths up to 89 mm (3½ in) and lengths of from 150 mm (6 in) to 300 mm (12 in). Thicknesses vary from about 25 mm (1 in) to 38 mm (1½ in).

They are generally laid on a sand and cement screed, or on a concrete base and are secured by mastic or bitumen emulsion mixed with a rubber latex.

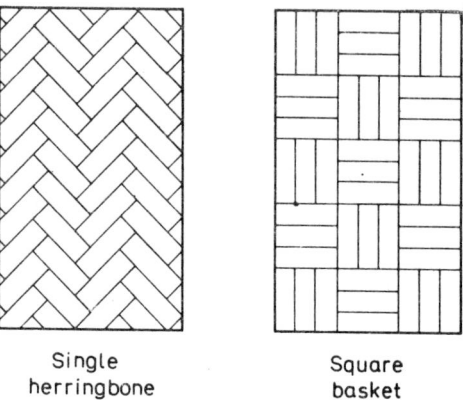

Single herringbone   Square basket

Fig. 2.2. Two typical designs for wood block flooring

Wood blocks can be laid in a number of patterns or designs. On very large floors sometimes two or more designs are used. Perhaps the most popular are single herringbone and square basket designs, illustrated in *Figure 2.2*.

Parquet and Mosaic Floors

Parquet floors have a most attractive appearance. They are made from specially selected decorative hardwoods, typical of which are oak, mahogany and teak. The pieces of parquet are produced in two main thicknesses, 6 mm ($\frac{1}{4}$ in) and 10 mm ($\frac{3}{8}$ in).

Parquet sections can be fixed individually to a smooth, even floor with glue or panel pins, or secured to a backing and laid as panels. When laid in this way panels of between 300 mm (12 in) and 600 mm (24 in) squares are normally used. The floors can be laid in a wide variety of designs to provide many decorative schemes.

Parquet strip, also known as overlay flooring, is very similar in appearance to hardwood strip flooring. The strips are normally of 10 mm ($\frac{3}{8}$ in) finished thickness, of widths from 50 mm (2 in) to 75 mm (3 in) and various, random lengths. The strips are normally tongued and grooved and end matched: they are usually secret nailed through the tongue and glued to the sub-floor.

WOOD GROUP OF FLOORS

Wood mosaic flooring is similar to parquet flooring except that the blocks of decorative hardwood are smaller, generally measuring about 113 mm (4½ in) long by 25 mm (1 in) wide. The depth of each block is usually about 10 mm (⅜ in). The blocks are normally formed into panels in various decorative designs and are fixed to the sub-floor by a suitable adhesive.

Because parquet sections, parquet strip and mosaic blocks are thin in depth, normally about 10 mm (⅜ in) thick, sanding should never be attempted. If sanding of worn floors is carried out the remaining life of the floor may be seriously reduced. In extreme cases the floor may be abraded away altogether. If treatment of this kind is contemplated consideration should be given to the use of abrasive nylon mesh discs, which will remove the top surface with minimum abrasion to the floor.

End-grain Paving

This paving, also known as end-grain block flooring, is used where heavy traffic conditions exist, for example in workshops and factories. The blocks are made from softwood, prepared in such a way that the end grain is vertical.

Blocks are normally 75 mm (3 in) by 225 mm (9 in) and from about 63 mm (2½ in) to 113 mm (4½ in) thick. They are usually laid on a concrete sub-floor and secured to the floor by means of pitch or grouted with bituminous compound.

Hardwood end-grain blocks, each about 25 mm (1 in) cube, are sometimes used as a decorative flooring in public buildings. They are normally laid on a screeded sub-floor.

CHARACTERISTICS

The physical qualities will depend not only on the type of wood, but also on how the wood has been cut. These qualities can vary widely, even with pieces cut from the same source. For example, timber which has been 'quarter-sawn' will resist abrasion and have greater durability than the same piece of timber 'flat-sawn'. Quarter-sawn timber is also more permeable to liquids than flat-sawn timber from the same source (*Figures 2.3* and *2.4*).

As wood is a naturally occurring material the varieties of grain, colour and decorative effect that can be obtained are almost infinite. An attractive appearance is, therefore, one of the most important features of a wood floor.

While colour has no bearing on the durability of a floor, it can affect maintenance and is important for this reason, as well as for its aesthetic effects.

The substance in wood that gives it colour is stable and fast to solvents, but soluble in water. Wood, therefore, loses its colour if treated with water over a period of time and so water must not be allowed to come

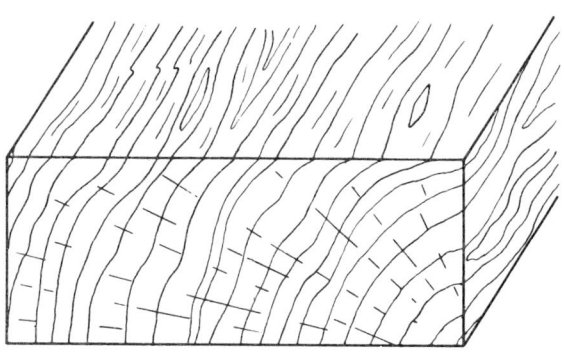

Fig. 2.3. Quarter-sawn timber

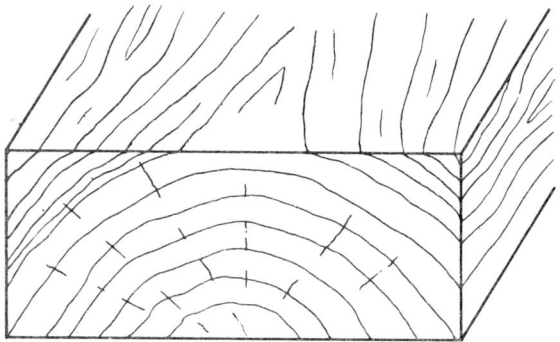

Fig. 2.4. Flat-sawn timber

into contact with wood. Prevention is best carried out by sealing the floor with a solvent-based seal and maintaining with a floor wax.

Colour can be affected to some extent by heat, but light can also have a marked effect. The effect of light varies greatly. Some timbers, for example afromosia, oak and iroko darken when exposed to sunlight. Others, such as gaboon and teak lose colour and fade. Maple, sapele and sycamore tend to become yellower on exposure to sunlight, while beech and western hemlock are examples of timbers that redden.

WOOD GROUP OF FLOORS

Acids, alkalis and a wide variety of chemicals can cause wood to lose colour. Wood affected by acids or alkalis should be treated immediately to neutralise any reaction that may have taken place. After neutralising, the floor should be well rinsed and allowed to dry. Brown discoloration caused by rust can be removed using a solution of oxalic acid in water, again followed by thorough rinsing.

It should be recognised that the colour of planed or sanded untreated wood is always different from that which has been treated with a solvent-based seal. It is generally accepted that all clear seals not only darken wood but tend to make it appear redder in colour. This is quite a normal phenomenon and is due to the seal's increasing the reflective properties of the wood surface. In many instances this change of colour, which brings out the natural beauty of wood, is highly desirable and is one of the main purposes of applying such a finish.

It has already been mentioned that wood can be adversely affected by water. Wood has a porous nature and readily absorbs water, either liquid or vapour. Similarly, in dry conditions water will evaporate from wood. This constant movement of moisture can cause wood to swell and shrink. Drying can cause gaps to appear in a wood floor. If, on the other hand, the moisture content rises above about 20 per cent, conditions can become such that the growth of fungi is encouraged and the floor may rot. Moisture content is, therefore, important and must be taken into consideration, particularly when sealing a new floor.

Wood stored in the open air normally has a moisture content of about 17 to 18 per cent. In an unheated building the moisture content averages between about 12 and 15 per cent and in a centrally heated building the moisture content is reduced to about 9 to 12 per cent. Where underfloor heating is present the figure could be as low as 6 to 8 per cent. If wood which has been stored in the open is not dried before it is used for flooring, it will lose moisture and shrink. Conversely if wood is over dried, it will absorb moisture, causing the floor to swell and lift. It is, therefore, always advisable to allow a new wood floor to reach its final moisture content, that is, to reach a state of equilibrium with the atmosphere, before sealing is attempted. This is particularly important with new wood block floors. If these are sealed before the wood reaches its final moisture content 'rafting' could occur. Rafting is the movement of a large number of blocks together at one time, causing a crack to appear in the floor. The crack can vary from about 3 mm ($\frac{1}{8}$ in) to about 25 mm (1 in) in width. This can occur if wood blocks are subjected to large changes in moisture content, causing them to shrink excessively, while at the same time they are tightly bonded together by means of the seal. Instead of swelling and shrinking individually into existing gaps, the blocks move as a mass, causing a large crack to appear.

WOOD GROUP OF FLOORS

The correct method of treatment for a new wood block floor is to apply a solvent-based liquid wax and maintain with a material of this type for about four months until moisture equilibrium is obtained, when the floor is sealed. Alternatively, the floor can be treated with a solvent-based liquid wax immediately after laying and before the initial sanding operation. Sanding then removes any wax from the surface, while allowing wax between the blocks to remain. Seal can then be applied overall. If shrinkage takes place, the wax between the blocks will enable each block to move individually, so preventing rafting.

These precautions are not necessary with new wood strip floors as the strips are usually secured by nails at each end, and so rafting cannot take place.

MAINTENANCE

General

The life of a wood floor, whether hardwood or softwood, depends not only on the correct selection of the wood for the traffic to which it will be subjected, but also on the maintenance given to the floor. Maintenance is extremely important. Correct maintenance can make a floor look attractive, facilitate cleaning and extend almost indefinitely the life of the floor.

Ideally, the maintenance procedure should be such that the floor is fully protected at all times from wear by traffic. In practice this is a very difficult standard to achieve and maintain, but nevertheless, the aim is realistic and maintenance schedules should regard this as one of the prime objectives.

Unsealed wood is porous and will readily absorb water and dirt. Untreated wood floors, therefore, rapidly deteriorate in appearance and become increasingly difficult to clean, as well as being unhygienic. Unsealed softwoods, particularly, wear quickly and become rough and detach splinters. By sealing, however, the poorer quality types of wood can be made satisfactory and even the best types are improved.

It has long been recognised that wood floors are improved by some form of surface treatment. Traditionally, three methods were used to fill the open pores at the surface and so protect the wood.

First, application of a vegetable or mineral oil was found to prevent water and stains from entering the floor, so improving the appearance. The oil also reinforced the surface fibres of the wood, with the result that the wearing qualities were improved.

This method, however, suffers from the disadvantage of the oil tending to hold any dust alighting on it. The surface is, therefore, almost permanently dirty and poor in appearance. A virtue of this property is made by describing the material as a 'dust allaying oil'. The oil is sometimes mixed with a hard resin to give a surface with good resistance to slip. These blends are known as gymnasium oil, as they were primarily used in gymnasia.

The requirement of floor oil has now practically disappeared because of the development in comparatively recent times of better floor sealing materials. Floor oil is, however, occasionally used in areas where absence of dust is vital and where the appearance of the floor is of little importance.

Second, waxes were used to seal the floor and are still used in considerable quantities even today. Waxes are, essentially, soft materials which penetrate into the open spaces and provide a surface which will prevent water, stains and other materials from entering the floor. They have the advantage over oil in that they retain their appearance and, when buffed, provide an attractive surface. They are not, however, hard materials and have poor durability compared with conventional floor seals. Even very light traffic can mark waxed floors, or remove the layer of wax. Waxed floors can be easily renovated by maintenance, but this is not always practicable or desirable.

Waxes have less resistance to slip than floor seals. They have poor resistance to stains and limited resistance to cold and hot water.

Wood floors subject to heavy traffic need the more permanent seals, followed by maintenance with wax, rather than wax alone.

Third, as a later development from wax, 'button polish', made from shellac resin and methylated spirits, was used and is still used on occasions. Button polish gets its name from the appearance of the pieces of shellac resin, which resemble buttons. It is generally applied liberally on wood as an initial treatment and a thin second coat applied after the first is dry.

The polish seals the open pores of the wood and supplies a surface which will resist penetration by water, and, to some extent, abrasion. Button polish needs to be protected with wax, as it scratches readily and quickly loses its initial good appearance, and is also easily stained by hot water and solvents. It is, however, very quick to dry as it does so purely by the evaporation of the methylated spirits component.

With the increasing demand for more durable finishes, work on oleoresinous seals produced a material that was not only better than oil, wax or button polish, but which could be easily applied and recoated when required.

Later developments for use on wood include the urea-formaldehydes, both one- and two-pot and the polyurethanes of various types. It is generally considered that the polyurethanes are the best types of seal for wood available at the present time, from the aspects of durability, speed of drying and chemical resistance.

Sealing Wood Floors

Almost all wood floors require sealing for protective, hygienic and decorative reasons. Sealing avoids the possibility of having soilage become embedded in the surface, resulting in difficult removal. The natural beauty of the wood is emphasised and the general appearance is considerably improved.

The range of solvent-based floor seals available has already been described in Chapter 1. The method of preparation for sealing depends on the type of wood floor and its previous method of maintenance, if any.

It should be recognised that while most of the different types of wood can be sealed perfectly easily and satisfactorily, some, by their nature, require more care if success is to be assured.

Perhaps the most important in this category are teak, keruing, gurjun and possibly Douglas fir. Teak has an oily nature and the remainder have a high resin content, so that resin exudation may continue even after a seal has been applied, particularly if the wood is new.

These timbers, therefore, should always be washed with solvent prior to sealing to remove as much oil or resin as possible.

Because of the wide variations that can occur in pieces of timber, floor seals do not always penetrate to the same extent. It is, therefore, essential always to apply the number of coats recommended by the seal manufacturer to ensure an even finish.

New Floors

A new floor, with the exception of parquet, should be sanded to remove any imperfections and to level its surface, then finished with a fine grade paper. Sanding should always be carried out by an experienced operator, as a floor can easily be damaged by unskilled operators. After sanding, the floor should be vacuumed or damp-mopped to remove any dust that may have been raised by the sanding process. Brushing will merely raise the dust to allow it to settle again later.

If the new wood floor is made of wood block or parquet, extra precautions must be taken if a plastic seal is to be used. The precautions are to prevent the possibility of blocks becoming stuck together with seal, with subsequent rafting leaving unsightly gaps in the floor.

A new wood block or parquet floor should first be treated with a thin application of a free-flowing liquid solvent wax over the whole area of floor, so that the wax will penetrate between the blocks, remain there and subsequently act as a lubricant. This will enable each block to move a fraction individually on shrinking, and so eliminate the possibility of rafting. Once the floor wax is dry the floor should be sanded to remove all traces of wax and any imperfections from the surface.

New wood strip and board floors can be sealed immediately after they are laid. No preliminary waxing is necessary before the sanding operation is carried out. This is because the strips or boards are held in position by nails, instead of the bitumen compound used to secure wood blocks to the sub-floor. Because the floor is nailed, rafting cannot occur with strip and board floors, even though slight shrinkage due to loss of moisture vapour may take place.

Old Floors

Ideally, old floors should always be stripped of any wax that may be present using the appropriate material, then sanded prior to sealing. The aim of sanding is to remove all dirt and previous surface treatments, including seals, which may prevent the new seal from adhering properly to the wood. Seals are formulated especially for application to the floor itself. If a seal is applied on top of another material the adhesion, and therefore durability, of the seal may be impaired. It should also be recognised that different types of seal may be incompatible. For example, the strong solvents in some plastic seals will soften, penetrate and lift oleo-resinous seals. The weak solvents in oleo-resinous seals will not be sufficiently strong to soften some plastic seals, with the result that an application of an oleo-resinous seal will not adhere to the plastic seal.

Sanding is, therefore, recommended if at all possible to enable the new seal to be applied directly to the wood itself, thus giving the best possible conditions for durability.

A word of warning, however, regarding the sanding of old wood floors, particularly those treated with an oleo-resinous seal. If the operation is not properly carried out the use of floor sanding machines can present a fire risk. Fires have been caused by ignorance, rather than negligence.

In each case, investigations have shown that the fire started several hours after all concerned with the sanding operation had finished work and left the premises. The cause was found to be due to the cloth bag of the sanding machine which was full of sawdust and fine particles of seal. High temperature caused by friction had started the process of combustion, and oxygen in the air in the bag had assisted the process. If the bag had been emptied or work had continued until the bag was full no oxygen would have been present and no fire would have started.

It is, therefore, always advisable at the end of any sanding operation to empty the bag and remove the contents to some place of safety, thereby eliminating any possible fire risk.

If sanding is not possible for any reason, the way in which the surface should be prepared will depend on the methods by which the floor has been maintained.

If the floor has been treated with button polish or has been only lightly sealed, the surface can be effectively prepared by stripping using an abrasive nylon mesh disc, grit 80 is recommended, under an electric polishing/scrubbing machine. These discs give results similar to those obtained by a very light sanding operation. After stripping, all dust should be removed from the surface with mopping equipment and the floor allowed to dry thoroughly before sealing. If mopping is carried out using solvent, for example white spirit (turpentine substitute), the solvent will evaporate quickly so that there is a minimum of delay between the mopping and sealing operations. If, however, water is used, the drying time will be longer due to its slower evaporation rate. It is essential that no water be present on the floor when seal is applied. Water may cause bubbles to appear in the seal or may turn the seal a milky colour, necessitating resanding and sealing.

If the floor has been maintained with a solvent-based wax, either paste or liquid, all traces of wax must be removed before seal is applied. This is absolutely essential, because no seal will adhere to wax. Removal of wax can be carried out with a solvent-based detergent wax remover. The material should be applied to the floor and allowed to soak into the wax layer for a few minutes, then scrubbed with a metal or nylon web pad under an electric polishing/scrubbing machine. The loosened wax can be removed with mopping equipment. It is important to change the water in the buckets frequently to be certain that the wax is completely removed from the floor and not just redeposited elsewhere.

Two applications of detergent wax remover may be required to ensure that as much as possible of the wax is removed from the floor. It is always recommended that after the wax stripping operation a test area should be treated with seal to be satisfied that all wax has, in fact, been removed. The test should be carried out on an area of about one or two square yards, preferably on the area of floor cleaned last. This is because if any wax is still present on the floor it is most likely to be on the area cleaned last.

To test, the seal should be applied and the drying time and hardness checked. If the seal is hard dry within the manufacturer's specified period it is reasonable to assume that all wax has been removed and the whole floor can be sealed. If the seal does not dry hard in the specified time, or dries patchily, it is possible that all wax has not been properly removed and the de-waxing operation should be repeated, followed by a further test.

WOOD GROUP OF FLOORS

If the floor has been maintained with a water emulsion floor wax the wax should be removed with an alkaline detergent. The weakest solution necessary to effect complete removal should be used, because strong detergent might require an excessive amount of rinsing. Once the floor is clean it should be rinsed thoroughly to ensure that no alkali remains on the surface. As an extra precaution it is advisable to add a little neutralising solution, or vinegar, to the rinse water so that any remaining alkali is neutralised. Failure to rinse thoroughly may result in the seal being affected by the alkaline detergent with consequent darkening in colour.

Waxing Wood Floors

It has already been mentioned in Chapter 1 that regular application of a floor wax will extend the life considerably and maintain the appearance of any floor seal. Therefore, it is always recommended that a sealed wood floor should be maintained by regular waxing.

Assuming that a wood floor has been sealed with an appropriate solvent-based seal, whether a solvent-based or water emulsion floor wax is used is largely a matter of personal preference. It is, of course, recognised that water should normally be avoided for regular cleaning of wood floors because of the detrimental effect of water on wood. A good seal, however, correctly applied, should be impervious to water and no water should come in contact with the wood itself. If a wood floor is well sealed, then a water emulsion floor wax can be used quite satisfactorily.

Sealed wood floors treated with a water emulsion floor wax must be kept under constant observation and inspected regularly, particularly in entrances and heavy traffic areas, for signs that the seal is wearing. As soon as there is evidence of wear, the wax should be removed and a further coat of seal applied to the worn area, followed by further maintenance with the emulsion floor wax. It must be appreciated that if the seal is not quickly repaired water will penetrate any worn spots into the wood itself. The presence of water can quickly cause wood to discolour and the entry of dirt into the floor itself may make cleaning very difficult. Furthermore, water may travel underneath seal in the area of the break and cause the seal to blister and detach from the surface.

Although the use of a water emulsion floor wax is perfectly satisfactory on a well sealed floor, a solvent-based wax is preferred. This is because if the seal wears to such an extent that floor wax comes in contact with the wood itself, no damage is caused to the floor and there is no possibility of the seal lifting in areas adjacent to the break.

If, however, for some reason it is decided not to seal the floor and to maintain with a floor wax, then a solvent-based floor wax only must be used. Water emulsion floor waxes are not suitable for use on unsealed wood floors and must not be used.

It should be stressed that maintenance of a wood floor by waxing only, without any seal present, should only be contemplated if there is unlikely to be heavy, wet or dirty traffic and then only if regular and frequent maintenance is possible.

In these cases, after the initial sanding operation, a solvent-based paste wax should first be applied liberally and well rubbed into the wood. When dry, the floor wax should be buffed with an electric polishing/scrubbing machine. Buffing helps the wax to penetrate further into the open pores, hardens the surface thereby improving the slip-resistance qualities and provides a pleasing sheen. If necessary, a further coat of paste wax should be applied and the process repeated.

Once a satisfactory surface has been prepared further waxing should be carried out using a solvent-based liquid wax in place of paste wax. The high solvent content in a liquid wax enables the floor to be cleaned as wax is applied. Liquid wax spreads easily and by leaving only a thin film on the surface minimises the possibility of a build-up of wax occurring. When dry, the floor should be buffed to harden the surface and produce an attractive sheen.

Waxing with either a solvent or water-based floor wax will result in safe, non-slip surfaces providing a build-up of floor wax is not allowed to take place. Should this occur, wax should be removed with the appropriate detergent and the maintenance cycle repeated.

Use of Detergents

If the floor has been sealed and waxed the type of detergent that should be used for routine maintenance depends on the wax used. For example, for a water emulsion floor wax, normal maintenance should be carried out with a dilute solution of a neutral detergent in water. Periodic stripping should be carried out with an alkaline detergent, ensuring that the floor is well rinsed with a neutralising solution before any further emulsion floor wax is applied.

Where the floor is maintained with a solvent-based wax, routine maintenance can again be carried out with a neutral detergent in water. Alternatively, a liquid solvent wax applied regularly will act as a cleansing agent and maintain the floor in a clean condition. Periodic wax stripping operations should be carried out with a solvent-based detergent wax remover.

Caustic materials should on no account be used on wood floors. These can cause permanent damage, both by destroying the natural colour of the wood and by causing wood to splinter.

WOOD GROUP OF FLOORS
WOOD COMPOSITION

During the 1940s there was an acute shortage of timber of all types in the United Kingdom and work was initiated to utilise to the full any waste wood that might be available. Wood chips from selected species of timber, industrial wood waste and forest waste, obtained from underdeveloped timber and thinnings from forest plantations and other sources, were processed into products known as wood composition or wood chipboard.

Wood composition rapidly became established as a flooring material in its own right, with certain advantages over wood floors. Unlike wood, wood composition can be manufactured to an exact specification, so that its physical properties can be accurately controlled. Defects, such as knots, sometimes found in wood are completely eliminated.

Wood composition is used extensively in domestic buildings and increasingly in industrial buildings, where it has withstood satisfactorily even heavy wheeled traffic on factory floors.

The basic raw material is wood. Wood composition therefore, consists essentially of cellulose and lignin, in the form of wood chips which are carefully graded for shape and size and are dried to a specified moisture content. Resins, often of the urea-formaldehyde type, are next mixed with the chips in predetermined quantities. The materials are then subjected to heat and pressure, when the loose mixture is compounded into a board.

After cooling, the boards are cut to size with saws before being conditioned prior to further treatment or dispatch to customers.

Wood composition is manufactured in various sizes, some boards or panels being supplied in tongue-and-groove form. Wood composition for flooring is commonly treated by the manufacturers with a finish to protect the surface and prevent dirt from being absorbed into the material, particularly when work is to be carried out by following tradesmen, for example painters and electricians. Precoating greatly facilitates cleaning.

The most common type of seal used to treat wood composition is based on polyurethane, although other types, such as oleo-resinous seals, are also used. Normally between one and three coats are applied during the manufacturing process and one or more further coats on site, after the floor is laid.

Wood composition is light in colour, the shade depending upon the species of timber used for the chips. If required, various shades can be obtained by the use of dyes. Water-soluble dyes are not recommended for this purpose.

Because there is no grain, wood composition will not warp or split when laid. Wood composition is warm and has good resilience. It can be covered with floor coverings, for example linoleum, flexible PVC, rubber and carpet. If uncovered, seal will provide a pleasing appearance and give a hard wearing, decorative finish.

Maintenance

It has already been mentioned that wood composition should be sealed with a solvent-based seal, preferably of the polyurethane type. Sealing is, indeed, essential if a floor covering is not to be used. This is because wood composition is very absorbent and will quickly deteriorate if water is allowed to penetrate the surface.

An uncovered floor should, therefore, be inspected regularly to ensure that the seal is intact. If any sign of wear appears, a further coat of seal should be applied before the seal becomes worn through.

If, however, the seal becomes worn and resealing necessary, the floor can be sanded to obtain an even surface, then sealed, as wood floors. Alternatively, abrasive nylon mesh discs under an electric polishing/scrubbing machine can be used to remove lightly any surface imperfections and prepare the floor for a new application of seal: 100 grit discs are recommended for this purpose.

It is always advisable to maintain the sealed floor with a solvent-based wax which will provide a pleasing appearance and prolong the life of the seal. While a water emulsion floor wax could be used if the seal is in excellent condition, a solvent wax is preferred. This is because if the seal does suffer wear, it is preferable for a solvent to come into contact with the wood composition rather than water. Water tends to swell the wood chips and can give the surface an uneven, bitty appearance, as well as remove the colour.

Routine daily maintenance can be carried out by damp mopping, using a solution of neutral detergent in water. The mop should be wrung out until it is almost dry, so that the minimum amount of water comes into contact with the floor.

Alternatively, a solvent liquid wax can be used in conjunction with a floor polishing machine. Periodically all old wax should be removed with a solvent-based detergent wax remover. The floor should be well rinsed and allowed to dry, followed by a fresh application of solvent-based liquid wax.

CORK

While the use of cork dates back to Roman days, cork floors were introduced in relatively recent times. The source is the bark of the cork-oak

WOOD GROUP OF FLOORS

or cork-tree, mainly found in the western region of the Mediterranean, particularly Spain and Portugal.

Cork is normally produced for floors in tile form and is found in a wide variety of domestic and industrial areas, for example libraries, art galleries, offices, hospitals and schools. It is generally not found in areas subjected to heavy industrial traffic.

Cork is composed of millions of extremely small irregularly shaped cellulose cells, bound together by resin, and while the cell walls are thin, they are very strong.

Cork tiles are made by grinding the cork bark into very fine particles, which are compressed and heated. This treatment releases the natural resins in the cork and synthetic binders or other resins are added as required, so bonding the cork particles together. Originally the heating process took many hours, but today the time has been reduced to only a few minutes by the use of special ovens.

After heat treatment the tiles are cut to the required size, ranging from approximately 100 mm (4 in) squares to 900 mm (3 ft) by 300 mm (1 ft), the most common size being 300 mm (1 ft) square. Thicknesses range from approximately 3 mm ($\frac{1}{8}$ in) to 14 mm ($\frac{9}{16}$ in), with either tongued and grooved or straight edges. Tile edges should be unchipped and firm.

Tiles are laid on many types of sub-floor, the main requirement being that the sub-floor must be level and dry. Perhaps the most common sub-floors are wood and concrete. If necessary, a damp-proof layer should be used.

Where possible, cork tiles should be secured to the sub-floor with headless steel pins, driven below the tile surface. If the sub-floor is not suitable for steel pins an adhesive should be used, after which traffic should not be allowed on the floor for two or three days, to allow the adhesive to dry thoroughly.

Characteristics

Cork tiles are normally produced in three shades of brown; light, medium and dark. They lend themselves to many decorative effects and are frequently installed in alternating shades to obtain attractive patterns. They are the softest of the resilient floors, since about half of the volume occupied by cork is air; are very quiet to walk on and restful underfoot. They are also warm and have excellent thermal insulation properties.

Cork is practically odourless. There is no tendency for it to splinter as the cellular structure does not lend itself to this defect. There is also little tendency for cork tiles to warp, as the cells containing air are not connected together in any regular pattern. The tiles are manufactured in varying densities, ranging from low density, below 450 kg/m^3 (28 lb/ft^3), to extra heavy density, over 580 kg/m^3 (36 lb/ft^3). Density of the tiles will, to some extent, influence the life of the floor.

Cork tiles have limited resistance to pressure. Indentations caused by the legs of heavy furniture, for example, are likely to remain until they are removed by sanding or other abrasive methods. Load spreading devices, such as domes of silence or furniture glides, should always be used on cork floors where there is a concentration of weight. Stiletto heels can ruin a floor in a comparatively short time. A floor subjected to stiletto heels should be treated with a very flexible seal, such as a two-pot polyurethane seal, so that it bends and assumes the contour of the heel mark without being punctured. If the seal becomes punctured water and dirt can penetrate and rapidly discolour the cork floor, as well as perhaps causing the seal to lift.

Although cork is physically unaffected by moisture and is very resistant to decay, it is, however, advisable to ensure that the minimum of water comes into contact with the floor, because water may seep between the tiles, causing them to lift.

Cork floors have a high degree of resistance to slip. Cork tiles are, therefore, very suitable for use on inclined areas or slopes and under rugs and mats.

Maintenance

While cork tile is not difficult to maintain if correct materials and methods are used, it should be recognised that if not properly maintained difficulties can occur. This is because untreated cork is extremely porous and will readily absorb oil, dirt and stains. The degree of resistance to dirt pick-up and wear by traffic depends, largely, on the extent to which the open pores in the cork are sealed and on subsequent maintenance. It is, therefore, essential that the open pores are sealed and that dirt is not allowed to penetrate into the tiles.

The initial method of maintenance will depend to some extent upon the original finish on the cork tiles. In general, four main categories of finish are recognised.

First, natural or untreated cork, which consists simply of ground cork manufactured into tiles and to which no resin or wax is added. Second, cork tiles treated during the manufacturing stage with a wax finish. Third, cork tiles strengthened with a resin wax, which appear smoother and are less porous. Fourth, cork tiles pretreated with a type of vinyl polymer which provides a surface impervious to stains and prevents ingraining of dirt.

Sealing Cork Floors

It has already been mentioned that cork should be sealed for decorative purposes to prolong the life of the floor and to facilitate subsequent maintenance.

Whichever type of seal is used, adequate allowance should be made for the porosity of cork. At least three coats of a solvent-based seal will probably be required to ensure that an even finish is obtained. The coverage will also most likely be less with the first coat than with subsequent coats and this should be taken into account when considering quantities of seal required.

The procedure to adopt with regard to sealing will depend on whether the floor is new or old and on the type of finish, if any, on the cork tiles.

New Floors

Initially, new cork floors should not be washed with water for four or five days after laying. This is to enable any adhesive to harden thoroughly. The tiles can, however, be lightly cleaned if necessary with the absolute minimum of a neutral detergent solution in water. If mopping equipment is used the mop should be only just damp and care should be taken to ensure that no water is allowed to seep into the joints between the tiles.

If the tiles are untreated they should be sanded, if necessary, to a smooth level surface, then sealed with a solvent-based seal. For the best possible results a two-pot polyurethane seal is recommended.

If the cork tiles have been pretreated with a wax finish, at the earliest convenient time all wax should be removed with a solvent-based detergent wax remover and the floor sealed with a solvent-based seal.

While cork tiles which have been treated during the manufacturing stage with a resin-wax can be maintained with a solvent wax only, it is preferable to use a solvent-based seal, if possible, followed by waxing.

If sealing is to be carried out, all resin-wax should be removed with a solvent-based detergent wax remover, followed by a light sanding of the floor. As this type of tile is sometimes supplied in a lighter gauge, it is advisable to consider this factor before sealing is attempted.

As with tiles strengthened with a resin-wax, tiles pretreated with a vinyl polymer can be maintained with a solvent wax only. Sanding and sealing with a solvent-based seal, however, will prolong the life of the floor, particularly if the vinyl becomes worn.

Old Floors

Cork tiles which have become discoloured, stained or worn can be refinished, if required, either by sanding or by using an abrasive nylon mesh disc under an electric polishing/scrubbing machine. Grit 80 is recommended, finishing with grit 100 or 120 to obtain a smooth surface.

Large areas can be treated quickly and easily with a drum sander. On cork, this should be used only by an experienced operator, because cork is soft and it is important that the depth of cut and evenness of sanding be controlled exactly. A drum sander in the hands of an inexperienced operator can cause unsightly lines to appear in the floor and can result in the removal of too much of the surface from the tiles.

Abrasive nylon mesh discs, however, can be used by operators experienced in handling a single brush floor polishing/scrubbing machine. Much less of the surface is removed if the machine is carefully handled and this method is preferred.

After removing the old surface the floor should be vacuumed to take up all the dust. Water should not be applied to the newly sanded floor as the cork tiles will be very porous. Once cleared of all dust seal can be applied.

Waxing Cork Floors

If, for some reason, it is decided not to seal new cork tiles, they should first be treated with a solvent-based paste wax. The wax will fill the tiny open cells of the cork and prevent penetration by dirt and water.

Once a surface has been established with paste wax, regular maintenance should be continued with a solvent-based liquid wax, buffing at frequent intervals.

A sealed cork floor should be maintained either with a solvent-based liquid wax or a water emulsion floor wax. Regular maintenance with a floor wax will keep the floor looking attractive and prolong the life of the seal.

While an unsealed cork floor should not be maintained with a water emulsion floor wax, treatment with a material of this type is perfectly satisfactory for a floor that is well sealed. This is because water remains on the surface of the seal and does not come into contact with the cork tile itself. If a water emulsion floor wax is used, the floor should be inspected regularly and frequently to ensure that the seal is intact, particularly in doorways, entrances and other areas subjected to heavy traffic. If the seal becomes worn, the floor wax should be removed with an alkaline detergent and further coats of seal applied, as necessary. Maintenance can then continue with the water emulsion floor wax.

Use of Detergents

When using detergents on cork floors it is important always to use the mildest material that will achieve the desired results. Neutral detergents can be used for routine maintenance if the floor is sealed. If mopping equipment is employed the mops should be barely damp, to ensure that the minimum amount of water comes into contact with the floor.

Strong alkaline detergents and harsh abrasives should never be used. While alkaline detergents do not produce an immediate deterioration in the appearance of cork, their continued use will reduce the life of cork tiles significantly. Dilute alkaline detergents can be used occasionally for the periodic stripping of old films of water emulsion floor wax.

Acid materials should be avoided, as acids will harm cork tiles.

If there are badly soiled or marked areas which cannot be cleaned by normal methods, the areas should be scoured with fine grade metal fibre or nylon web pads and the dust removed with a vacuum cleaner. Coarse grades of metal fibre and nylon web pads should not be used as they may leave marks on the surface of the cork tiles.

When using floor polishing/scrubbing machines, just enough action to remove the dirt should be employed. By allowing the machine to linger in a particular area any finish which may be present could be damaged.

If oil impregnated mops are used, it is important that no deposits of oil or other materials be allowed to remain on the floor. Oil is easily absorbed and can cause unsightly dark stains.

When properly laid and maintained, cork tiles will give many years of quiet, luxurious service.

MAGNESITE (MAGNESIUM OXYCHLORIDE)

Magnesite is known by a variety of names, including magnesium oxychloride, oxychloride composition, magnesite jointless composition, magnesite composition, jointless flooring, or sometimes just oxychloride or composition.

Magnesite floors are laid *in situ* and often require a screed over the sub-floor to provide a level surface. The sub-floor can be concrete, stone, or occasionally wood.

One advantage of magnesite flooring is that it can be laid over an existing rough floor, or even over old paving stones and worn boards. One, two or three coats can be applied to almost any thickness to obtain a flat, even surface. While magnesite floors are generally laid in jointless form, different coloured squares can be laid, if required, by means of wood dividing strips.

Such floors are used for industrial, commercial and domestic purposes. They are particularly serviceable in kitchens and similar areas where resistance to oil and grease is important.

WOOD GROUP OF FLOORS

Magnesite is formed by the result of a chemical process, its installation being similar to that of concrete. It consists of a number of ingredients, perhaps the most important being calcined magnesite in finely divided powder form. The calcined magnesite consists essentially of magnesium oxide and this is mixed with a solution in water of magnesium chloride, when a chemical reaction takes place. The substance formed has good cementing properties and provides a distinctive type of floor.

Fillers include wood flour, or sawdust, with a moisture content of not greater than 15 per cent. Other fillers and aggregates include such materials as ground silica, talc or powdered asbestos. The main requirement of any filler used is that it shall be inert and nonreactive.

The proportions in which the various fillers and aggregates are included determine the type of floor that will be produced. Hard, strong floors are formed by using a high proportion of fillers and aggregates of mineral origin. A more resilient, softer and quieter type can be produced by introducing increased proportions of wood flour or sawdust. In the United Kingdom magnesite floors generally contain a significant proportion of wood flour or sawdust and are, therefore, very sensitive to moisture.

Pigments are included to provide colour, and plasticisers are also sometimes present.

The ingredients are thoroughly mixed together, then trowelled to the required thickness and smoothness. The thickness may vary from 13 mm ($\frac{1}{2}$ in) for a single coat to several centimetres for two or more coats and will depend upon the evenness of the sub-floor and the durability required. After four to six hours it is given a final hand trowelling.

Characteristics

In appearance, magnesite has a surface similar to that of smooth cement. Magnesite floors are produced in a variety of colours, perhaps the most common being a dark red shade, given by a red oxide pigment, and black. It can be laid in a single, plain colour, or marbled.

Old magnesite floors are sometimes found to suffer from a marked loss of colour, resulting in a drab, grey appearance. This is due to normal traffic abrasion. Colour can, however, easily be restored using either a pigmented floor seal or pigmented paste wax.

Magnesite floors are generally rather hard, but not as hard as cement or concrete. They have a tendency to be noisy and are comparatively warm underfoot.

Resistance to abrasion and impact are good, as also are the wearing properties. Magnesite is unaffected by mineral oils, fats and greases. Since magnesia, the main component of such floors, is an alkaline material, acids must not be allowed on the floor. Strong acids will tend to dissolve the floor and mild acids may scar the floor permanently. Strong alkalis are also injurious and must be avoided.

Magnesite can become slippery if not properly maintained.

One of the special characteristics of magnesite is that it is very sensitive to moisture, because of the presence of wood flour or sawdust filler in the mix. The floors are hygroscopic and if unsealed and washed frequently with water may develop cracks, or, in extreme cases, lift from the substrate. This is because the wood flour or sawdust will absorb water and swell. On drying, the floor shrinks again, causing stress. Repeated cycles of swelling followed by shrinking can cause the floor to crack and deteriorate rapidly.

Maintenance

Once the floor has set hard it is advisable to cover it with a material which will prevent dirt from being trampled into the surface and at the same time allow water vapour to evaporate from the magnesite. Suitable materials are absorbent paper or sawdust.

If possible, the floor should be allowed to harden for at least three days before being opened to traffic. The length of time required for it to harden is dependent, to some extent, on the atmospheric temperature. Hardening takes place quicker in warm than in cold conditions. If the temperature is cold it may be advisable to allow a longer period before allowing the floor to be used.

Even after the floor has dried, the hardening process will continue for several weeks, and because of this, abnormally heavy traffic should not be allowed during this period. It is also advisable during the hardening period to prevent the floor from being exposed to extremes of heat or cold. If possible, it should be guarded from the direct rays of the sun and cold draughts.

Perhaps the most important single factor to be considered in the maintenance of magnesite floors is the highly absorbent nature of the wood flour or sawdust filler. Washing or scrubbing should, therefore, be avoided and as little water as possible allowed to come into direct contact with the floor. Continuous, regular washing will denature the floor as water, particularly when used with alkaline materials, affects the oxychloride cement and exposes the filler to the injurious effects of water.

Sweeping powders containing an excess of hygroscopic salts, such as calcium chloride, should also not be used on magnesite floors.

WOOD GROUP OF FLOORS
Sealing Magnesite Floors

It is the general practice of many flooring contractors to seal magnesite floors immediately after they have been laid. If, however, a new floor has not been sealed it should be treated with a solvent-based seal as soon as it has hardened thoroughly. Sealing will improve the appearance, prevent water, dirt and stains from penetrating into the floor and provide a surface for subsequent maintenance.

Old floors which have not previously been sealed should be cleaned to remove all traces of any wax which may be present, as described in the section on wood floors. Because of the wood content in magnesite, old floors can be sanded if required. Alternatively, the top surface can be removed with abrasive nylon mesh discs underneath a floor polishing/ scrubbing machine. All dust should then be removed. When the floor is thoroughly clean and dry a test area should be treated prior to sealing the whole floor, to ensure that the desired results are obtained.

When the sealed surface eventually shows signs of wear, it is recommended that a further coat of seal be applied to restore the appearance and, perhaps more important, to prevent the surface from dusting.

Waxing Magnesite Floors

As magnesite is unaffected by solvents of the white spirit type, waxing with a solvent-based wax is recommended. If the magnesite has not been sealed, application of solvent wax will suit the slightly absorbent nature of the magnesite. The wax will fill any tiny open pores and prevent penetration by dirt and water.

Regular maintenance should be carried out with a solvent-based liquid wax, rather than a paste wax. Liquid wax will both clean and wax the floor in a single operation. Buffing at regular and frequent intervals will maintain the floor in a clean, hygienic and attractive condition.

A well-sealed magnesite floor can be maintained either with a solvent-based wax or a water emulsion floor wax. If a water emulsion floor wax is preferred, a coloured material may be used, as the presence of colour in the floor wax will help keep the floor looking attractive.

Periodically all old floor wax should be removed with the appropriate detergent, the floor rinsed, using the minimum of water, and allowed to dry. If a mild alkaline detergent is used it is always advisable to add a little vinegar, or neutralising solution, to the rinsing water. After allowing the floor to dry maintenance should then continue with the appropriate floor wax.

WOOD GROUP OF FLOORS

Regular waxing of a magnesite floor will provide a pleasing appearance, greatly simplify cleaning and prolong the life of the floor.

Use of Detergents

It has already been mentioned that magnesia, the main component of magnesite floors, is alkaline and is harmed by acids. Similarly, the use of strong alkalis and harsh abrasives should be avoided. Some milder alkalis may also prove injurious and should not be used. Not only will they remove any water emulsion floor wax which may be present, but they could also expose the wood filler to the detrimental effects of water.

Repeated use of a strong alkaline detergent may cause white patches to appear on the floor. Perhaps the best remedy in this event is to treat the floor with a pigmented seal. This will provide an even colour and facilitate subsequent maintenance.

Routine daily maintenance should be carried out by sweeping, if necessary, followed by washing with a solution of a neutral detergent in water. Only the minimum amount of water should be allowed to come into contact with the floor and any mops should be wrung out almost dry.

If there are badly soiled or marked areas which cannot be cleaned with a neutral detergent solution, they should be scoured with fine grade metal fibre or nylon web pads under an industrial floor polishing/scrubbing machine, in conjunction with the minimum amount of detergent. Coarse grades of metal fibre and nylon web pads should not be used as they might leave scars on the surface of the magnesite.

Correct maintenance procedures, together with the use of approved materials, will ensure the best possible service from magnesite floors.

3

STONE GROUP OF FLOORS

Included among the stone group are a wide variety of different types of floor of both manmade and natural origin.

Perhaps the most important are concrete and granolithic floors and these are considered first. Then follow terrazzo and a separate section on marble, a natural stone. Terrazzo and marble are very similar in their characteristics and methods of maintenance. Other natural stones are considered together in the next section and these include granite, limestone, sandstone, quartzite and slate.

Clay, or ceramic, tiles include a wide variety of different types. Considered in this section are mosaic, quarry, faience, paver, vitreous, tessellated, adamantine and encaustic tiles and industrial pavoirs.

A small, separate section is devoted to brick and the final section deals with cement latex.

CONCRETE AND GRANOLITHIC

The term concrete is generally applied to a range of floor materials including various types of cement.

Compositions of limestone and clay were used in Roman times as cements for buildings. Although there have been many refinements in techniques since those days, the basic ingredients have changed little.

Concrete is generally laid in heavy duty areas. Its ability to sustain considerable weights has made it the most popular type of floor for industrial undertakings, particularly where machinery is installed or heavy traffic is present. It is generally recognised as a strictly utilitarian type of floor and is comparatively low in cost.

The specification for concrete, including the grades of cement and aggregate used, will depend on the intended use of the floor, whether for industrial or domestic purposes.

Concrete is formed by dispersing coarse aggregates in a fine paste of cement and water, well mixed together. Cement is manufactured under controlled conditions. Its composition is specified to produce a uniform material. The range of cements available is very wide and includes, for example, high alumina for quick setting.

STONE GROUP OF FLOORS

Limestones, consisting mainly of calcium carbonate, are burned, or calcined as this process is often termed. The burning operation forms quicklime with the evolution of carbon dioxide. The quicklime is slaked and when the cement is mixed into a mortar and spread it reabsorbs carbon dioxide from the air to re-form calcium carbonate, which sets to a hard mass.

The aggregate can vary in size to form different grades of concrete. Sand, a siliceous material obtained from rock, is often known as fine aggregate. Pebbles, gravel or other inert materials that will unite with cement are referred to as coarse aggregates.

Once the concrete has been laid, a screed, consisting of a mixture of sand, cement and water, is often used as a finish. Both the concrete and the screed can be made to include various materials, such as carborundum or a colour pigment, to meet the special requirements of any particular floor.

Concrete is generally laid in one or two courses, depending on the type of finish required. In thickness, the concrete floor may range up to eight inches in depth.

The term granolithic is used in connection with a particular type of surface and the method employed in its manufacture. Granolithic floors mostly have a normal concrete sub-floor with a topping generally including granite chippings, to give a less porous and more dense surface. Although the normal aggregate is granite, other crushed, hard materials can be used to provide a satisfactory wearing surface.

Granolithic concrete is probably most widely employed for factory and other heavy duty industrial floors. When properly laid it will give excellent service even under heavy traffic conditions.

Granolithic floors generally require a minimum of about seven days to dry hard and longer may be required in cold weather. During the hardening period the floor should be kept in a damp condition and heavy traffic should not be allowed.

Granolithic floors are for the most part drab in appearance, but coloured pigments can be added, if required. They have a very low porosity, the porosity of granite being less than 1 per cent, compared with about 10 to 18 per cent for limestone and 10 to 12 per cent for brick. The presence of granite also increases resistance to impact and to abrasion. When well laid, granolithic floors are very slip resistant.

If a granolithic floor is not dried properly, signs of cement dusting could appear after a few months. It is also noisy and cold and, in general, rather difficult to keep clean.

The method of laying will depend on the area. While a small area of approximately 20 m^2 (20 yd^2) would normally be laid in one operation, a larger area would generally be laid in bays.

Concrete is sometimes produced in slab form, often of about 0·2 m^2

(2 ft^2). The advantage here is that worn slabs can be removed and replaced when required.

Characteristics

Plain concrete flooring is unattractive in appearance and very drab if left untreated. A pigment added to the mix greatly improves the appearance, by giving a coloured finish. Various shades of red, brown, cream and green can be obtained at slight additional cost. However, the range of pigments that can be added is limited by the alkaline conditions during the manufacturing process and later in the finished concrete.

One of the advantages of a coloured floor is that glare, which can be extremely unpleasant under certain lighting conditions, is reduced. Also, a red colour can give a warmer appearance to the concrete.

Concrete is extremely hard and noisy. The concrete made from Portland cement is resistant to a wide range of chemicals, including mineral oils and grease, but it is slowly attacked and broken down by acids. Vegetable oils, fats and sugar solutions are also harmful.

Most concrete floors have good resistance to slip, but this, however, will depend almost entirely on the surface finish. If a concrete floor becomes slippery, roughening the surface by mechanical means is not advised since it could result in dusting, thereby creating further difficulties. Etching the surface with acid has also been tried, but has generally not proved successful and any tendency to reduce slipperiness has been very short lived. Where slipperiness is caused by a smooth, hard aggregate becoming exposed, etching with acid may well remove more cement, exposing more aggregate and making the floor even worse.

If slippery conditions occur, the floor should be treated with a seal, as seals are generally non-slip.

All types of concrete, including tiles and granolithic, show a tendency to dust, some more than others. Dust can greatly increase cleaning costs, particularly in areas where it is liable to come into contact with precision engineering machines. Dust can be harmful to machinery, assembly parts, finished components in stores and during the packing operation. Loose dust particles are abrasive and constant traffic can grind them into the floor, causing still more dust. In extreme cases dust can become a health hazard. It is, therefore, essential for reasons of both economy and hygiene that concrete floors should be well maintained.

It is sometimes difficult to ascertain whether dust on a concrete floor originates from the concrete or from working processes. A test can, however, be carried out quite simply by collecting some dust from the floor and adding a few drops of hydrochloric acid. If the dust originates from the concrete, small bubbles of carbon dioxide gas will appear on the surface of the dust. This is because the acid will react with the lime present in the concrete dust according to the formula:

$$CaCO_3 + 2HCl \rightarrow CaCl_2 + H_2O + CO_2$$

Limestone Hydrochloric Carbon dioxide
acid gas

Maintenance

All concrete floors are porous to varying degrees. Unprotected floors often start suffering the effects of abrasion, evidenced by dusting, as soon as traffic is allowed on them.

Of the many possible causes of deterioration, perhaps the most important are spillage of chemicals and exposure to oil and grease.

These problems can be overcome by sealing the floor, although it should be recognised that best results will be obtained from sealing the floor only if the concrete is of reasonably good quality. Surfaces that are weak or friable cannot be treated effectively.

Sealing Concrete and Granolithic Floors

Selection of the correct type of seal is most important. Concrete contains alkaline materials and, when newly laid, the lime salts are very active. If a conventional paint is applied the presence of these salts will prevent drying and subsequent adhesion. Also, if such products are applied to concrete after it has hardened, the combined effects of alkali and moisture remaining in the concrete will cause the varnish constituent to saponify, with the result that the paint film will break down and detach from the floor.

Special seals are, therefore, required which will penetrate into the concrete, bind the surface dust particles together and provide a clean, durable surface. Seals will also facilitate maintenance and, when pigmented, provide a pleasing appearance.

Of the clear types of material available for sealing the floor, perhaps the best known are those based on sodium or potassium silicate, often known as silicate dressings. Other materials, such as those based on zinc or magnesium fluorosilicate are also sometimes used.

When a dilute solution of sodium silicate is applied it penetrates into the concrete and chemically reacts with it to form calcium silicate, an insoluble, glass-like material. The surface becomes extremely hard, with the result that the problem of dusting is overcome, or minimised, and resistance to water, oil and acids is greatly increased.

Silicate dressings can be applied to new concrete after about 7 to 14 days, when it has hardened sufficiently to take light traffic. Old concrete can be treated provided the surface is thoroughly clean.

A silicate dressing is comparatively cheap and very large areas can be treated economically. It is very easy to apply, perhaps the best method being through the rose of a watering can. Two or three coats are generally applied, allowing 12 to 14 hours drying time between each application.

When the dressing shows signs of wear a further coat should be applied on the worn area. No special preparation is required, other than to ensure that the floor is thoroughly clean and free from all traces of oil, wax and grease.

If, however, it is intended to seal the floor with a pigmented seal at a later date, application of a silicate dressing is not advised. This is because this dressing will make the concrete more difficult to etch and render the surface less permeable. Penetration of seal could, therefore, be impaired, resulting in loss of adhesion and durability.

While a silicate dressing will have very little effect on the appearance of a concrete floor, a pleasing colour and a new look can be given by using a pigmented seal. These are widely used to improve the appearance and to strengthen the surface layer and prevent it from dusting. Some pigmented seals, for example those based on polyurethane, are also highly chemically resistant. Application of seal also provides a smooth surface which is easy to maintain and keep clean.

Of the many types of pigmented seal available, including those based on phenolic and epoxy resins, perhaps the most important and widely used are those based on synthetic rubber and polyurethane resins.

Before applying any seal it is extremely important that the floor be clean. Particular attention should be given to the removal of any oil, grease or wax that may be present to enable the seal to key properly to the concrete. If the concrete is smooth it should first be etched to achieve the best possible adhesion and durability.

Prior to etching, all loose concrete and dust should be removed by brushing or other means. Several different materials are satisfactory for etching, those in most common use perhaps being sodium bisulphate, hydrochloric acid, also known as muriatic acid and spirits of salts, and phosphoric acid.

If the first is used, the floor should be dampened with water and the sodium bisulphate sprinkled over the surface at an approximate spreading rate of 25 m^2/kg (14 yd^2/lb) weight and allowed to dissolve. During this process the floor is etched, a visible sign being the appearance of small bubbles of carbon dioxide gas on its surface. Once the reaction has ceased, the floor should be rinsed thoroughly to remove all traces and allowed to dry.

If hydrochloric acid is used, a 10 to 20 per cent solution in water should be prepared by pouring the acid into the water. Concentrated phosphoric acid should be diluted to make a 10 per cent solution. Acid should always be added to water, because if the latter is poured into acid bubbling sometimes occurs and harmful acid may be thrown into the operator's face or over his hands. Rubber or plastic gloves are recommended and plastic buckets are preferred, as metal buckets can be attacked by acid.

The floor should be dampened with water and the etching solution spread evenly over the surface, at an approximate spreading rate of $1 \cdot 5 - 2 \text{ m}^2/\text{l}$. ($50-75 \text{ ft}^2/\text{gal}$). An old mop should be used as it will be destroyed by the etching solution. The solution should be allowed to work for about 15 min, until the effervescing has ceased. Since this is the last step in the cleaning operation before sealing, several rinses are recommended. Failure to remove all traces of acid could have an adverse effect on the adhesion of any seal subsequently applied. After rinsing, the floor should be allowed to dry thoroughly for at least overnight.

The following day the concrete should be tested for the presence of moisture. One method of testing is to lay a solid rubber mat of at least $0 \cdot 1 \text{ m}^2$ (1 ft^2) on the floor, and after about three hours inspect the under-surface for the presence of moisture. If moisture is present, the concrete should be allowed to dry for a longer period of time.

If the floor has been properly etched it will have the appearance and texture of fine grade sandpaper. In some instances a second etching may be necessary. Thorough rinsing should follow each etching process.

Should the floor have been previously sealed, all old seal should be removed, if possible, with a chemical stripping compound or sanding machine. If complete removal of the old seal is not possible, it is imperative to ensure that the new seal will be compatible with the old one, otherwise the new seal may either lift the old seal or not adhere to it, resulting in poor durability.

Having ascertained that the seals are compatible, the floor should be thoroughly clean and dry before the new seal is applied.

Once the floor is prepared, seal can be applied with a lambswool bonnet, roller applicator or brush. For rough concrete, a Turk's head brush may be used.

A seal based on synthetic rubber resins, supplied in one-pot (ready for use) form will give a pleasing appearance, has good durability and will resist many forms of chemical attack, including oil and grease. It is easy to apply and can be readily renewed or touched-up when required. Two or three coats are generally recommended, each coat being allowed to harden overnight.

Of the polyurethane pigmented seals, perhaps the best are those supplied as two-pot materials. This type is superior in all respects to those based on synthetic rubber resins, except for simplicity and, perhaps, recoating properties. Two-pot polyurethane pigmented seals give excellent appearance, resistance to chemicals and durability. Three coats are generally recommended, each coat drying in about two hours, suitable for overcoating. The final coat should be allowed to harden overnight before traffic is allowed on the floor.

Waxing Concrete and Granolithic Floors

Only on very rare occasions will it be necessary to obtain a polished surface on concrete floors. It should, however, be recognised that a good quality floor wax will provide a barrier against scuffing by foot and other traffic and considerably extend the life of a floor seal. While either a solvent or water emulsion floor wax can be used on sealed concrete, a water emulsion floor wax is preferred. It is the only type that can be used if certain synthetic rubber floor seals are used, as some seals of this type are softened by white spirit and similar solvents. Water emulsion floor waxes also have a greater resistance to slip, often an important factor on hard concrete and granolithic floors.

Use of Detergents

A concrete or granolithic floor that has been sealed is comparatively easy to keep clean. The amount and type of soil will influence the selection of the detergent to be used. In general, routine cleaning can be carried out effectively with a neutral or mildly alkaline detergent in water. The floor should then be rinsed and allowed to dry.

If, however, the floor has not been sealed, dirt, oil and grease can easily penetrate and discolour it. Excessive scrubbing with strong alkaline detergents should be avoided, as they can weaken the cement binder in the concrete and expose the aggregate, thereby producing a rough surface which will be more difficult to clean.

Soap also should be avoided, as it will react with the lime in concrete forming a scum, which will cause the floor to resoil more rapidly.

Many concrete and granolithic floors suffer from oil and grease stains, particularly in garages, mills and workshops. Volatile, flammable solvents such as white spirit should not be used to remove these stains. Oil and grease are normally freely miscible with solvents of the white spirit type and the solvent may cause them to penetrate further into the concrete.

Oil and grease stains can be removed effectively with a degreasant, which may be either water or solvent-based. Of the water types, detergent crystals based on sodium metasilicate are extremely effective for cleaning large areas at a comparatively low cost. The detergent crystals can be used by two methods. If hot water is available, the crystals are added to the water and dissolved, and the solution spread over the area to be cleaned and allowed to soak into the soil. The floor is scrubbed either with an electric polishing/scrubbing machine or a deck scrub, then rinsed and allowed to dry.

Alternatively, if hot water is not available, the area may be dampened with water and the detergent crystals sprinkled on to it, when they will slowly dissolve in the water. After soaking overnight the area should be scrubbed, then rinsed.

If the soil has penetrated into the concrete a second application of detergent crystals may be necessary, as soiled patches may appear after about 12–18 hours. The soiled patches are due to dirt being drawn to the surface by the action of the detergent crystals.

Small areas of concrete contaminated by oil and grease can be cleaned very effectively with a solvent-based detergent wax remover. This type of material consists of solvent and water blended with emulsifiers and other additives. The material is spread over the affected area and allowed to soak for a few minutes, then scrubbed with an electric polishing/scrubbing machine or deck scrub.

The solvent and emulsifiers lift the oil and dirt from the floor and enable it to be removed with mopping equipment and water.

If the floors are extremely dirty, with many layers of dirt, wax, oil or grease impacted on them, steam cleaning using a fortified alkaline detergent may be required. Alternatively, dirt can be loosened with scarifying brushes under a scrubbing machine. Loose dirt can be swept up by brushing and any remaining can be removed by using detergent crystals or a solvent-based detergent wax remover. Once all traces of dirt have been removed, the floor should be rinsed thoroughly and sealed.

TERRAZZO

Terrazzo floors have been in existence for a very long time, and several in the Mediterranean area are more than 3 000 years old. One terrazzo floor was found in excellent condition at Pompeii, where it had lain covered in lava and volcanic ash for many centuries.

These floors have grown rapidly in importance with the large number of buildings erected in recent years. They are found in entrance halls, office corridors, hospital operating theatres, lavatories and washrooms and as the main traffic aisles in large departmental stores.

Terrazzo is frequently laid where a high standard of appearance and cleanliness are required. It consists, essentially, of crushed marble aggregate embedded in a cement matrix. The marble aggregate is normally angular, as distinct from elongated and flaky, and is carefully graded by size, depending on the effect required. The marble can be of many different colours, which contribute to the attractive, decorative effect. The cement may be white or coloured.

Terrazzo is generally laid on a concrete base over which a screed has been applied. It is normally laid in one of two forms, *in situ* or precast, the latter sometimes being known as terrazzo tiles.

Where it is laid *in situ* it is usually divided into panels about $1 \cdot 2$ m^2 (12 ft^2) in area separated by strips, which may consist of plastic, brass, aluminium alloy or ebonite.

Terrazzo tiles are made from the same materials as *in situ* terrazzo, but are cast in moulds and are normally more expensive, but their quality may be more reliable. The most common size of tile is $0 \cdot 1$ m^2 (1 ft^2).

After the floor has been laid for about four days it is ground, to produce a smooth even surface, with a coarse abrasive stone and a plentiful supply of water. Final polishing is generally carried out by machine, using carborundum as the abrasive.

Characteristics

The vast majority of terrazzo floors are extremely attractive in appearance, particularly where many different coloured marble chips are used.

The floors tend to be hard, noisy and cold, but have extremely good wearing properties and will withstand heavy foot traffic. They are resistant to water and can stand up to any amount of it used for cleaning the floor.

Terrazzo floors are not, however, resistant to acids. These attack both the marble aggregate and cement grouting, causing them to become pitted. Even mild acids have a detrimental effect and must be avoided.

Some harsh alkaline powder materials are also detrimental to terrazzo. When a floor is mopped with a solution of such a material in water, the solution penetrates the minute open cells of the terrazzo. As the water evaporates the residual powder recrystallises and expands, exerting considerable pressure on the surrounding material. The result is that the surface texture disintegrates into a fine dust. Continual treatment of this sort will cause the floor eventually to become pitted and retain dirt, making cleaning difficult.

If the surface has been damaged by either acid or constant use of harsh alkaline powder materials the surface can be restored to its original smoothness by honing.

Terrazzo floors generally have good resistance to slip, particularly when polished to a fine finish. If, however, the floor is too highly polished by the use of very fine carborundum stones or other special polishing aids, a slippery surface may result.

Maintenance

It has already been mentioned that acids and harsh alkaline powder materials should not be used on terrazzo floors.

Oils are also detrimental to terrazzo; once having penetrated the surface they can be extremely difficult to remove. Oily types of mop dressing and sweeping compound should, therefore, be avoided.

If, however, oil has been allowed to penetrate into a terrazzo floor, it can be removed with either degreasing agents, such as detergent crystals, or a poultice of whiting and white spirit, mixed together and applied thickly over the area. More white spirit should be added periodically and the poultice allowed to remain overnight before being removed. A further poultice may be required if the oil has not been absorbed by the first application.

A poultice, compounded of whiting, sodium citrate and other additives, is also used to remove rust stains from terrazzo floors. These are sometimes caused by metal objects which have been allowed to remain on the floor for a period of time, or by small metal particles from steel wool or metal fibre floor pads. The latter should never be used on terrazzo. Nylon web pads only should be used.

Some disinfectants, particularly those based on phenols and cresols, may take part in a chemical reaction with small amounts of iron compounds present in the cement if used as cleaning agents. The normal result of such a reaction would be a pink discoloration which may persist, even if the surface is reground.

Sealing Terrazzo Floors

The advisability of sealing terrazzo floors with solvent-based seals has long been debated. While some tung-oil oleo-resinous seals have been advised in the past, they are apt to have a pronounced yellowing effect resulting in discoloration of the floor.

Even the lightest coloured solvent-based seals are also liable to suffer from the same defect, which can badly affect the appearance of white and pale coloured terrazzo floors. Thought should also be given to difficulties that will occur when the seal starts to show signs of wear in the main traffic lanes. Touching-up seal on terrazzo floors to ensure an even appearance is extremely difficult.

Terrazzo can vary widely in porosity, with the result that some seals cannot obtain an effective key to the floor, resulting in loss of adhesion and poor durability.

On balance, therefore, a solvent-based seal is not recommended. If, however, a seal is required, particularly if the floor is slightly open and pitted, a water-based seal of the acrylic polymer resin type should be used. Water-based seals are water-white in colour and do not discolour on ageing. This is of great importance when white, or very light coloured floors are being treated.

The application of one or two coats of water-based seal will prevent dirt from entering the floor and greatly facilitate the routine maintenance operations.

Waxing Terrazzo Floors

A floor wax is an excellent maintenance material for terrazzo. Discrimination in selection is, however, essential. Solvent-based wax, either paste or liquid, has a tendency to darken terrazzo floors and may cause slippery conditions, especially if the floor is wet. Their use is not, therefore, recommended.

Water emulsion floor waxes, particularly those based on light coloured acrylic polymers, are very suitable for use on terrazzo. They not only maintain the floor in an attractive condition but also assist in providing a non-slip surface and make the floor much easier to clean and keep clean.

Use of Detergents

It is inevitable that over a period of time the floor will slowly become dirty and a gradual discoloration may well escape casual notice. For this reason spot cleaning should be carried out periodically, using a mild abrasive paste on a piece of damp cloth or nylon web pad.

When a thorough cleaning is necessary either a mild alkaline liquid detergent or a mild abrasive paste or powder should be used, followed by thorough rinsing. The detrimental effects of harsh alkaline powder materials have already been mentioned and these compounds should not be used.

Daily routine cleaning should be carried out with a solution of a neutral detergent in water, in conjunction with either a floor polishing/scrubbing machine or mopping equipment.

The use of soap on terrazzo is not recommended as it is liable to result in a slippery surface, particularly if a build-up of residue is allowed to take place. Such a build-up can be removed with a mild abrasive powder and water, in conjunction with a polishing/scrubbing machine and nylon web pads.

Terrazzo is easy to maintain with a water-based seal, water emulsion floor wax and neutral detergent solution, and is highly decorative and durable.

MARBLE

Marble floors have been known for several thousand years. The many that remain in excellent condition even today testify to marble's durability. While a few marble floors wear out, many are spoilt and discoloured by faulty maintenance.

Like terrazzo, marble is used where a high standard of cleanliness and appearance are required. Entrance halls and corridors of public buildings are probably the two main areas where it is found. It is an expensive flooring material, although the cost can be reduced by using small pieces as a marble mosaic.

Marble paving is normally about 19 mm ($\frac{3}{4}$ in) to 25 mm (1 in) thick and can be laid in squares or rectangles or to any design. The marble slabs are laid on a bed of mortar; any mortar brought to the surface at the joints should be removed by wiping with a damp cloth as soon as possible after laying.

Marble is a naturally occurring stone, included in the class of rocks known as limestone. In general terms, marble can be defined as any limestone hard enough to take a polish. Many grades and colours are found. Perhaps the best known of the marbles suitable for flooring, either as slabs or mosaic, are: travertine, which is straw-coloured; Belgian black, which is very hard-wearing; Roman stone, covering a number of attractive, colourful marbles; Sicilian, the hardest of the white marbles; and Ashburton, which is not geologically a true marble.

After laying, marble is generally honed to provide a lustrous finish, which tends to hide scratches. Highly polished marble is normally not used for floors, but for walls or other surfaces not liable to abrasive wear.

Characteristics

Marble is impressive in appearance and extremely attractive. It is, however, mostly very hard and tends to be noisy and cold. Marble floors are

STONE GROUP OF FLOORS

generally extremely durable, but one or two of the slightly softer types of marble may wear quicker than the harder types, causing an uneven appearance. Marble is resistant to water but, like terrazzo, is readily attacked by even dilute acids which can etch and roughen the surface. Strong acids could cause a hole to appear in the floor, or destroy the grouting altogether.

Harsh alkaline powder materials are also detrimental to marble and by crystallising underneath the surface could cause it to disintegrate into dust. While honing can restore a damaged surface, it is a task for specialists.

Marble usually has good resistance to slip although it can become slippery if wet, but resistance can mostly be improved by application of a suitable water emulsion floor wax.

Maintenance

The principles concerning maintenance are the same as those for terrazzo floors. Acids and harsh alkaline powders, oily compounds and soap must be avoided. Disinfectants also are not recommended. If a seal is required, a water-based seal of the acrylic polymer resin type should be applied. This should then be maintained with a water emulsion floor wax. Routine maintenance should be carried out with a solution of neutral detergent in water, together with mopping equipment or a floor polishing/scrubbing machine. If a machine is used, the pads should be of the nylon web type and not metal fibre, as splinters may detach and cause rust marks to appear on the surface of the marble floor.

In general, a marble floor requires little maintenance. While neglect will spoil its appearance, it will not normally impair its serviceability. When properly maintained, it is doubtful whether marble floors can be excelled by any other type for brilliance and timeless beauty.

NATURAL STONE

Among the many types of natural stone used for flooring perhaps the most important are granite, coarser varieties of limestone, sandstone, quartzite and slate. While marble is also a naturally occurring material it has been considered in a separate section.

Natural stones are generally laid in slabs which vary in area and thickness according to the type of stone and requirement. Some are used for flagstones, which can be made from any rock which can be easily split along the natural bedding plane.

STONE GROUP OF FLOORS

Natural stones are widely found in churches, public buildings, institutions, markets, farmhouses, dairies and other areas, particularly where there is heavy foot traffic. They are usually laid on a screeded concrete sub-floor and bedded in a mortar or cement grout, which appears at the joints between the stone slabs.

Characteristics

Granite originates from igneous materials that have cooled slowly into extremely hard, compact masses. It has a crystalline structure and varies widely in colour and appearance. Colours range from white and silver to dark grey, with occasional shades of green, blue and black. If required, granite can be polished with abrasive material to a high gloss finish.

Limestone consists almost entirely of calcium carbonate, with some impurities. It is generally light grey or parchment in colour, with occasional cream, yellow and pale brown shades.

One of the better known is Ashburton limestone, which is grey with red and white streaks and resembles marble in appearance. Other well known limestones include Enstone, creamy in appearance with a characteristically rough surface and Horton rag and blue rag, which are essentially blue, sometimes with brown markings.

Sandstone is formed from sedimentary deposits and consists almost entirely of sand grains. Colours are typically red/brown, but some sandstones are pale yellow with an occasional grey or blue shade.

Quartzite, like sandstone, is formed originally from sedimentary deposits of sand grains. These were cemented together with silica and then transformed, under tremendous pressure during earth movements, into one of the hardest rocks known. Layers of sediment contain minute flakes of mica, which appear along the cleavage planes. The mica reflects light, so that the surface appears to sparkle in a most attractive manner. Colours range from silvery grey-green to olive and gold shades.

Slate is a metamorphic rock, produced as a result of argillaceous clay and shale deposits being subjected to great heat and pressure by earth movement. Colours are generally sombre and include blue-grey, various shades of green and blue-green. Slate has a very fine texture, not typical of other natural stones.

All the natural stones discussed are hard, quartzite being extremely hard. Sandstone is, perhaps, the softest.

All are noisy underfoot and cold. Resistance to abrasion is very good and all are extremely hard wearing.

The chemical resistance of natural stone depends on its composition. Granite will resist alkalis, acids, oils and water. Limestone will resist alkalis and water but is damaged by acids and stained by oils and grease.

Sandstone has good resistance to water but can be damaged by strong alkalis. Sandstone is also stained by oil and grease.

Quartzite is virtually unaffected by acids, alkalis and other chemicals in general use. It is one of the most chemically resistant of the natural stones. Slate is also practically immune to all common chemicals. It is completely resistant to water and not affected by alkaline detergents.

Granite is inclined to wear smooth and to be slippery in worn areas. Limestone, sandstone and quartzite are not slippery; the latter has slip resistant properties even when wet. Slate also has very good slip resistant properties, when both dry and wet. This is probably because it has a slightly irregular surface and is not completely smooth.

With regard to special properties, granite has an extremely low porosity, generally less than 1 per cent.

Maintenance

Immediately after laying it is advisable to remove any grout or cement bedding material which may have appeared on the surface of the stone. This can best be carried out by hand using a damp cloth. If, however, the grout or cement has dried hard, it can be removed from those stones that resist acid, for example granite, sandstone, quartzite and slate, with a solution of hydrochloric acid diluted in the proportion 1 part acid to 4 parts water. A small area should be treated at a time, removing the cement with mops as soon as it is loosened.

It is important that this operation be carried out as quickly as possible because acid will attack the cement between the stone slabs. After cleaning, the floor should be rinsed very thoroughly, several times, to ensure that no acid remains. It is advisable to use an old mop and have the acid solution in a plastic rather than in a metal bucket. Operators should wear rubber gloves.

The maintenance of a stone floor mostly depends on two main factors, namely the type of stone and the porous cement or grout between the stones. While stone itself is relatively easy to maintain, it must be recognised that porous cement or grout can be affected by acids and many other chemicals. Acids and strong alkaline detergents should, therefore, be avoided. Harsh abrasive materials can damage the cement or grout and may even scratch soft stone. If an abrasive is required for any purpose, a test should first be carried out on a small area in an out-of-the-way place to ensure that it is suitable for the task.

It has already been mentioned that some stones are stained by oil and grease, resulting in dark unsightly discoloration. Oily type sweeping compounds and oily dust mop dressings should, therefore, be avoided.

If stone has become discoloured with oil it can be removed with detergent crystals, as described in the section dealing with concrete on page 44. A poultice of whiting and white spirit can also be used, if required.

Steel wool should not be used on stone floors because of the possibility of fragments becoming detached and subsequently rusting. Nylon web pads are preferred.

Sealing Natural Stone Floors

In general, it should not be necessary to seal natural stone floors. In any event solvent-based seals should be avoided because of the possibility of subsequent yellowing of the stone, and also because of the difficulty of obtaining a satisfactory key to the surface. For the most part, a solvent-based seal will not penetrate into stone sufficiently to obtain the required degree of adhesion. While a solvent-based seal can, therefore, be applied, it will probably wear off in a comparatively short period of time.

It is, however, recognised that on rare occasions a seal may be necessary, perhaps because the natural stone has worn or become pitted and porous. Alternatively, a seal may be necessary to protect the cement or grout, particularly if this constitutes a comparatively large area of floor. On these occasions, and only when absolutely necessary, a water-based seal can be applied. This should be of the acrylic type, practically water-white in colour and non-yellowing on ageing. Application of one or two coats will render the floor easier to maintain in a clean and hygienic condition.

Waxing Natural Stone Floors

In normal circumstances a floor wax is unnecessary on natural stone floors. The use of a solvent-based wax, either paste or liquid, may result in slippery conditions and should, therefore, be avoided.

On occasions a water emulsion floor wax may be used either to improve the appearance, or to assist in providing a non-slip surface, particularly if the natural stone has worn smooth. Granite is an example of a floor that is sometimes rendered more non-slip by application of a water emulsion floor wax, but in general its use is not recommended.

Use of Detergents

Natural stone floors should normally be maintained by dry sweeping to remove all loose dirt and soilage, followed by mopping with a solution of a neutral detergent in water. The floor should then be rinsed and allowed to dry.

If necessary, granite can be treated with a fine scouring powder to clean stubborn dirt from the surface.

Only non-corrosive materials should be used on limestone; acid materials attack it and must not be used. If sandstone is allowed to become very dirty and a stronger material than a neutral detergent is required, it can be cleaned with a mild abrasive in conjunction with water.

If they become very dirty, both quartzite and slate can be cleaned with a solution of an alkaline detergent in water. After scrubbing, the floor should be thoroughly rinsed, adding a little neutralising solution or vinegar to the rinse water. Under normal traffic conditions, however, routine maintenance should be carried out with a solution of neutral detergent in water.

CLAY (CERAMIC) TILES

Clay tiles include all types of floor tile having a basic clay, or argillaceous, composition, and range from cheap quarry to the more expensive vitreous tiles.

The craft of ceramic tile manufacture is extremely old and its beginnings are lost in antiquity. The Egyptians were using Nile mud to make tiles over 6 000 years ago and were applying glaze finishes several thousand years before the Romans laid tiled floors in Britain. That many of these floors are in perfect condition even today is evidence, if any is needed, of their outstanding durability.

Mosaic glazed and unglazed tiles and quarry tiles are today manufactured by almost the same methods as those used by the ancient Egyptians. Clay, with some water added, is placed in a mould and heated, or 'fired', at temperatures of about 1000 to 1200°C (1832 to 2192°F) for several days in a kiln. During this process some complex chemical reactions take place. The atmospheric conditions in the kiln are very carefully controlled and firing is probably the most critical stage in the manufacturing process.

Developments in comparatively recent times have resulted in a number of different types of tile being produced. Perhaps the main ones for floors are two classes of tile known as floor quarries and floor tiles. While the distinction between the two is not precise; floor tiles mostly have a finer finish, whereas floor quarries are made with wider dimensional tolerances from coarser clay. Floor quarries, or quarry tiles as they are generally known, normally have an unglazed surface.

STONE GROUP OF FLOORS

The word quarry comes from the French word 'carreau', meaning a square or paving tile. Quarry tiles are normally about 150 mm (6 in) square and of 13 mm ($\frac{1}{2}$ in) thickness, although many other sizes are available.

Quarry tiles are very hygienic and used extensively in food factories, breweries, dairies, kitchens and canteens. They are also widely used in washroom areas and in lavatories. They are manufactured with a wide range of surface. While the majority are plain and smooth, other surfaces are available to improve slip resistance qualities, for example ribbed and chamfered, or chequered squares of various sizes.

A further selection of tiles for floors include glazed tiles, mosaics, faience, paver, vitreous, tessellated, adamantine and encaustic tiles, and industrial pavoirs.

Glazed tiles are frequently slippery when wet and for this reason are seldom used for floors, being normally employed to decorate walls. Moreover they can also chip and crack if subjected to sudden impact and the glaze can wear under constant foot traffic.

The word mosaic is often used in connection with floor tiles. The term, however, generally refers to a pattern or picture formed by laying different coloured pieces of tile, stone or glass, side by side. In the main, therefore, mosaic refers to design rather than to a type of tile. Pieces of ceramic tile used for mosaic often have small irregular shapes.

Faience tiles are decorative and may be either glazed or unglazed. They may be laid in a single colour, alternating colours or in mosaic patterns.

Paver tiles are unglazed and resemble quarry tiles in appearance, but are made by the dust-pressed method instead of being extruded. They are heavy duty tiles, particularly suitable for areas subject to heavy traffic conditions.

Vitreous tiles are manufactured from refined clays, mixed with added flints and felspars, at kiln temperatures above $1200°C$ ($2192°F$). The word vitreous means resembling glass, and at these high temperatures the mix fuses to form a glass-like material. These tiles are extremely durable and are frequently laid for decorative effects in many bright colours. Carborundum is sometimes added to give improved slip resistance qualities.

Tessellated tiles resemble mosaic and are generally not as strong as plain quarry tiles. They are normally made in squares of from 25 mm (1 in) to 150 mm (6 in). They are highly abrasion resistant and are used under heavy foot traffic conditions, for example canteens, kitchens and school cloakrooms.

Adamantine tiles resemble quarry tiles but are manufactured to a more precise specification. They are designed to withstand heavy traffic, including such items as metal-wheeled trucks. They are generally from 13 mm ($\frac{1}{2}$ in) to 25 mm (1 in) thick and are found in breweries, factories, workshops and similar areas subject to heavy traffic conditions.

Encaustic tiles are inlaid with different coloured clays to form a pattern and are generally less resistant to abrasion by foot traffic than are plain tiles. They are not normally laid where heavy traffic is expected as they are primarily intended for their very pleasing decorative effect.

Industrial pavoirs are manufactured by dry-pressing powdered clay or extruding clay in a plastic state before cutting, drying and firing. They are normally supplied in rectangular shapes up to 50 mm (2 in) thick and are intended for heavy industrial applications. Colours include yellow, red, brown, cream and white. They have various surface textures and may include silicon carbide to provide a slip resistant finish, or they may be ribbed or chequered, depending on the requirement. Tiles may also have a non-ferrous metal included in their manufacture. These tiles are particularly abrasion and chemically resistant and are used in dairies, breweries, refineries and similar locations.

Floor tiles are usually laid on a concrete sub-floor with a sand and cement screed. As the tiles are laid mortar will extrude to the surface along the joints between the tiles. After laying, the joints are topped up with a sand and cement grout until the level in the joints is flush with the tile surface. Any surplus grouting material should be cleaned off immediately by covering with sand or clean sawdust and wiping with a dry cloth. This prevents subsequent staining of the tiles with the cement.

After laying, the tiles should be allowed to stand for at least four days before the surface is cleaned with water and traffic is allowed on the floor.

Characteristics

Modern quarry tiles are available in a range of colours, including blue, blue/grey, buff, brown and black. Perhaps the most commonly used, however, are tile-red in shade. They are generally very hard, noisy underfoot and cold, but have outstanding resistance to abrasion and are exceptionally durable.

Where quarry tiles are liable to be subjected to sudden impact by heavy objects, the denser and thicker types are preferred. Thin tiles may chip at the edges where heavy impact is encountered.

They are very resistant to grease, oil and a very wide variety of chemicals, and are generally resistant to alkalis, and many types will also resist acids.

Quarry tiles can be slippery when wet or polished with a solvent-based wax. Special surfaces to improve slip resistance are available and reference has already been made to some of the types available. Sometimes an abrasive is included in the grout between the tiles to improve resistance to slip.

In general most types of quarry tile have a very low water absorption and excellent resistance to staining. They are completely vermin- and rot-proof.

Maintenance

Newly-laid quarry tiles are often stained with cement, giving an unsightly appearance. Cement can be removed with a solution of hydrochloric acid in water, as described in the section dealing with the maintenance of natural stone floors, page 52. Acid solutions, however, should not be used on glazed tiles as they sometimes tend to dull the surface.

Quarry and other types of clay tile require little in the way of routine maintenance and are comparatively easy to keep in a clean and hygienic condition. Many floors are maintained perfectly satisfactorily with water containing a little neutral detergent and mopping equipment.

While upkeep of the tiles themselves is easy, incorrect maintenance materials and methods can destroy the grout between the tiles. This can result in a rapid deterioration in appearance, and perhaps discolouring of the face of the tiles. They may also become loose and eventually detach from the floor. Grout can be attacked by acids and harsh alkaline detergents and they should, therefore, be avoided. Coarse abrasives may also damage the grout, and only mild abrasives should be used. But even these should not be used on glazed tiles as they may dull or scratch the smooth finish.

In recent years grouts based on epoxy and furan resins have been developed. These materials are superior in quality to conventional grout, but more expensive. Once grouts based on either epoxy or furan resin have thoroughly cured they are impervious to alkalis, greases, oils, detergents and many types of acid. They are not, however, widely used, and generally are specified only when required by special circumstances.

Oily types of sweeping compound and oily mop dressing may stain the grout and cause discoloration. The tiles also may become discoloured, particularly if they are unglazed. Oil may be removed with detergent crystals, as described on page 45.

If floor polishing/scrubbing machines are used on tile floors, only nylon web pads and not steel wool pads should be employed, as fragments left on the floor might rust and stain the tiles or grout. If rust stains appear, they can be removed with a poultice by the method described on page 47.

STONE GROUP OF FLOORS

Perhaps one of the most common cleaning problems is the removal of hard water deposits and soap scum from washrooms and shower floors. Both the deposits left by hard water and soap scum are alkaline, due to the lime content and cannot be removed with normal types of detergent. They can, however, be removed with a mild acid solution, such as hydrochloric acid in water. The solution should be allowed to act for a few minutes and the floor scrubbed, if necessary, with an old nylon web pad. All acid solution should then be thoroughly rinsed from the floor with clean, hot water. Rubber gloves should be worn and great care taken to protect the skin from splashing.

Mildew and growth of fungus sometimes occur in hot, moist areas; for example bathrooms and showers. These can be removed effectively by scrubbing with a detergent containing a quarternary ammonium compound, such as benzalkonium chloride. Regular application, approximately once every two to three weeks, will eliminate them completely. Products of this type are quite safe on the skin and are odourless and non-irritating.

Sealing Clay Tile Floors

In general, it should not be necessary to seal clay tile floors. Solvent-based seals should not be used because of the difficulties of obtaining satisfactory penetration into the tile and subsequent adhesion. While a solvent-based seal will dry satisfactorily on a tile floor, it might readily detach and flake off even under light traffic. It might also subsequently darken in shade and discolour the tiles.

A seal may, however, be required on rare occasions, perhaps to protect the grout if this comprises a comparatively large area of floor. One or two coats of a water-based seal of the acrylic type should be applied. The seal fills any tiny pores that may be present and resists the absorption of moisture or stains, so making the floor easier to maintain.

Waxing Clay Tile Floors

It is generally unnecessary to wax clay tile floors. Solvent-based wax, either paste or liquid, should be avoided. Solvent waxes could cause the floor to become slippery, particularly in washrooms and similar areas where water may be present on the floor. Pigmented paste waxes are sometimes used to restore colour to faded floors. These are generally satisfactory with regard to slip resistance as the pigment imparts slip retardent qualities. Over-application should, however, be avoided, as a build-up of floor wax could cause slippery conditions.

STONE GROUP OF FLOORS

If a floor wax is required to protect either the tiles or the grout, a water emulsion can be used. The latter will improve the appearance and protect the floor from traffic, as well as making routine maintenance easier.

A water emulsion floor wax can also be used to improve the slip resistance qualities of a floor surface that may have been worn smooth. In general, however, under normal conditions a floor wax is not recommended.

Use of Detergents

A clay tile floor should, for the most part, be maintained by sweeping to remove all loose dirt and soilage, followed by mopping with a solution of neutral detergent in water. It should then be rinsed with clean water and allowed to dry.

While soap and water have been used traditionally for cleaning clay tiles, if the floor is not thoroughly rinsed afterwards a thin film of soap scum may remain, resulting in slippery conditions. The thin film of scum can also attract and hold dirt, so that the appearance can deteriorate.

It has already been mentioned that acids and harsh alkaline detergents should not be used and that if an abrasive is required only mild types should be considered, and then not on glazed tiles.

If the tiles become very dirty they can be cleaned with a solution of a mild alkaline detergent in water. The floor can be scrubbed, if required, with a floor polishing/scrubbing machine and nylon web pads, followed by thorough rinsing.

BRICK

Bricks for flooring usually consist of clay, either vitreous or semi-vitreous. Sand-lime and concrete bricks are also sometimes used for flooring. Some bricks are manufactured with a chequered or ribbed surface to improve slip resistance qualities.

Bricks are laid either flat or on edge, generally on a sand and cement screed over a concrete sub-floor. Colours are normally rather drab, although 'blue engineering' bricks are sometimes used. They are noisy, cold to the tread, very abrasion resistant and hard wearing. Bricks have a particularly high resistance to compression and impact shock, especially when laid on edge.

The chemical resistance of bricks is very good, the critical factor being the grout between them. They can become slippery when wet.

STONE GROUP OF FLOORS

Normal maintenance should be carried out by sweeping all loose dirt and soilage, followed by washing with a solution of a neutral detergent in water. If the floor is particularly dirty it can be scrubbed with a scouring powder and warm water. Soap should not be used, as it tends to form a scum which could cause slippery conditions. This is particularly the case in hard water areas.

Sealing is not generally recommended, but should a seal be necessary one or two coats of a water-based seal of the acrylic type might be applied.

While brick floors are not normally waxed, if necessary a water emulsion floor wax can be used to improve the appearance and facilitate maintenance. Solvent-based waxes should not be used as they might cause the floor to become slippery, particularly when wet.

Ribbed and chequered brick surfaces may harbour dirt unless they are cleaned regularly. This should be carried out with a floor polishing/scrubbing machine or deck scrub, or hose and water.

CEMENT LATEX

Cement latex floors are laid *in situ* and consist of a mixture of cement, aggregate, fillers and pigment, to which is added an aqueous emulsion of latex.

The cement may be Portland or high alumina and the aggregate generally consists of crushed natural stone or sand, or pieces of vulcanised rubber or hardwood chips. Fillers may include asbestos powder, powdered slate, ground limestone and wood flour. Pigments can be added to provide a variety of colours. The latex normally consists of a dispersion in water of natural rubber, although proportions of bitumen emulsions or polyvinyl acetate emulsions are sometimes included, depending upon the requirements. Rubber-based types will resist water and alkalis and those consisting essentially of polyvinyl acetate will resist oil and grease. The floors are generally laid on a concrete base and are steel trowel finished. They are normally about 6 mm ($\frac{1}{4}$ in) thick.

Cement latex floors are commonly laid in industrial premises where some degree of chemical resistance is required, for example kitchens and dish-washing areas. They are not as hard as concrete floors and are quieter to the tread. Wearing properties depend on the proportion of latex used and on the type of aggregate. Vulcanised rubber and wood chips give more resilient floors, but if a high degree of resistance to abrasion is required stone aggregate is used. High alumina cement is employed when resistance to weak acids, vegetable oils and sugar solutions is required. Cement latex floors are generally slip resistant, even when wet. They are not normally subject to dusting.

STONE GROUP OF FLOORS

These floors are not difficult to maintain in a clean condition, but without regular maintenance they can become drab. Cleaning should consist of brushing to remove surface dust and soilage, followed by periodic mopping with a solution of neutral detergent in water. Strong alkaline detergents, acids and harsh abrasives should not be used. Scrubbing also should be avoided.

If required, the floor can be sealed with a water-based seal of the acrylic type. Two coats should normally be applied. Sealing will fill the open pores of the surface and prevent dirt and stains from entering the floor, as well as making routine maintenance easier. The seal should then be maintained with a water emulsion floor wax. Solvent-based waxes should not be used on cement latex, as solvent might damage the floor.

When properly maintained, cement latex floors will give excellent service over many years.

4

ASPHALT GROUP OF FLOORS

The asphalt group of floors includes mastic asphalt and pitch mastic. Because of their similarity in characteristics and methods of maintenance, they are considered together.

Asphalt floors are sometimes included among stone or hard floors, but while asphalt, like many types of stone, is a naturally-occurring material, it is not usually as hard.

Asphalt may be affected by climatic changes of temperature and is sometimes softened during the summer months when warm or hot conditions prevail. In this and some other respects, asphalt floors differ from stone floors.

Methods of maintenance are also different. For these reasons, therefore, the asphalt group is considered separately from the stone group of floors.

MASTIC ASPHALT AND PITCH-MASTIC

The use of mastic asphalt can be traced back to Babylonian days, when it was used as a waterproof membrane for floors. In more recent times, perhaps the most important advance was the discovery by Sir Walter Raleigh in 1595 of an asphalt lake in the island of Trinidad. A further 250 years passed, however, before it was developed commercially. Lake asphalt, as it is generally known, consists of bitumen blended with fine clay and silica.

Another type of naturally-occurring material is known as natural rock asphalt. It is found in France, Switzerland and Sicily and many other places throughout the world. It consists of calcareous rock impregnated with about 6 per cent of bitumen. Natural rock asphalt is mined or quarried. After processing it is generally used in mastic asphalt.

To allay any possible confusion that might arise over the various terms used in connection with mastic asphalt and pitch mastic floors, the following general descriptions can be applied.

Asphalt is taken to mean a blend of bitumen with a proportion of inert mineral material. It is often described by its origin, for example Trinidad lake asphalt.

ASPHALT GROUP OF FLOORS

Bitumen can be described as a solid or viscous liquid which softens when heated. It is obtained from petroleum and is also found as a natural deposit or as a component of asphalt.

Mastic asphalt comprises asphaltic cement and mineral aggregate, blended to form an impermeable mass. It is solid or semi-solid at normal temperatures, but becomes fluid and can be spread by a hand float when heated.

Asphaltic cement is a mixture of bitumen or lake asphalt or similar material with an oil, known as a flux oil, which is used to soften the bitumen and make it less viscous.

Pitch mastic consists of aggregates and a binder made up of coal tar pitch and a flux oil, to which lake asphalt may be added. The inclusion of lake asphalt makes the pitch mastic more resistant to cracking and indentation. The aggregate may be either fine or coarse and generally consists of graded limestone, siliceous or igneous rocks or stone fragments.

A pigment of ground metal oxide is sometimes added to give colour to the floor. Mastic asphalt and pitch mastic floors are laid *in situ*, generally on a concrete sub-floor. Asphalt is heated to about 200° to 230°C (392 to 446°F), when it melts and can be spread by wooden floats to the required thickness. Asphalt flooring is generally laid in one or two coats, and no traffic should be allowed on the floor until it has cooled to normal room temperature.

Light, medium and heavy-duty grades are available and thickness can vary from about 16 mm ($\frac{5}{8}$ in) to 19 mm ($\frac{3}{4}$ in) for domestic and office floors, and 25 mm (1 in) and over for heavy-duty factory floors and loading bays. Special grades of both mastic asphalt and pitch mastic are available to meet specific requirements, such as floors in areas where acids or oils are liable to be spilled. Special grades are also sometimes used in dairies and abbatoirs and in buildings where sparks must be avoided to prevent the risk of an explosion or fire, for example granaries and some ordnance factories.

In recent years rapid progress has been made in the development of various types of asphalt for specific conditions. Over 50 different grades are now available to satisfy a very wide variety of requirements.

Characteristics

Mastic asphalt floors are normally available in dark colours only, generally black, red or brown although occasionally a green or grey floor is laid. Pitch mastic floors are also made in a limited range of colours, normally black, red or brown.

Hardness varies widely according to the composition and requirements of the floor. While some floors are relatively soft, specially hardened asphalts are sometimes specified. The low melting point of the asphalt matrix can be a disadvantage under abnormally high temperature conditions. In these circumstances some grades of asphalt may show indentations where concentrated loads are placed on them, for example the thin legs of heavy furniture. Glides should always be placed under the legs of furniture and other heavy objects to spread the load over as wide an area as possible. Pitch mastic can become brittle at low temperatures.

Hard asphalt floors can be noisy and rather cold to the tread, although some of the softer grades are quieter and somewhat warmer.

The abrasion resistance and durability of both mastic asphalt and pitch mastic are extremely good and both types of floor can be laid to withstand heavy traffic.

Both mastic asphalt and pitch mastic have limited resistance to chemical attack. Normal grades are attacked by most acids, but special grades will withstand attack by dilute acids. Mastic asphalt will resist alkaline solutions at normal temperatures, but pitch mastic has only a very limited resistance to attack by alkali. Mastic asphalt is normally non-resistant to animal, vegetable and mineral oils, fats and greases. Pitch mastic is also susceptible to animal and vegetable oils, fats and greases, but will resist mineral oils and greases at normal temperatures.

Petrol, white spirit, paraffin and similar solvents will soften both mastic asphalt and pitch mastic floors. Use is sometimes made of this fact to remove any doubt of whether a particular floor is asphalt or magnesite. Asphalt and magnesite floors often look alike, particularly if worn or dirty. Identification can be extremely important as maintenance procedures for the two types of floor differ widely. A test can be carried out by cleaning a small area with a solution of neutral detergent in water, then pouring a little petrol or white spirit (turpentine substitute), on to a cloth and rubbing the cleaned area. If the floor is magnesite the solvent will have no effect, if asphalt the floor will soften slightly and the cloth will become discoloured.

Both mastic asphalt and pitch mastic have good resistance to slip, although the floors can become slippery when wet.

Mastic asphalt and pitch mastic are impervious to water, are vermin and rot-proof, odourless after laying and dustless.

Maintenance

While mastic asphalt and pitch mastic floors vary slightly in their composition and properties, for maintenance purposes it is convenient to treat them both in the same way under the general heading of asphalt.

An asphalt floor requires relatively little maintenance, although the use of correct materials is essential to obtain maximum service and the best decorative effects. Use of incorrect materials can cause considerable damage to an asphalt floor. Conventional solvent-based seals and solvent waxes must not be used. If sweeping compounds are employed they must be entirely free from mineral oil or other material detrimental to asphalt.

Heated objects should never be allowed to come into contact with the floor, as they may cause permanent disfigurement of the surface.

Normally, asphalt floors should be maintained with a water emulsion floor wax and a solution in water of a neutral or mild alkaline detergent. In certain circumstances, however, and providing special precautions are taken, it is possible to treat an asphalt floor with a solvent-based pigmented seal.

Sealing Asphalt Floors

It has already been mentioned that a conventional solvent-based seal should not be used because the solvents could soften and damage the asphalt. If a seal is required, a water-based seal of the acrylic polymer resin type, preferably coloured either red, brown or black to match the colour of the floor, should be applied.

New asphalt floors generally have a surface film. Before applying seal it is advisable to remove this film by scrubbing with an alkaline detergent solution in water. The floor should then be well rinsed and allowed to dry. If the surface film is not removed, the floor seal will 'crawl' (like water on grease), and dry with a spotty uneven finish.

Old asphalt floors tend to lose their original colour under constant traffic and may appear faded and drab. Application of a coloured water-based seal will restore lost colour to the floor, thereby improving the appearance.

One or two coats of water-based seal should be applied, followed by maintenance with a water emulsion floor wax.

While conventional solvent-based seals should not be used on asphalt, it is recognised that on rare occasions, perhaps to improve the chemical resistance of the floor or to restore colour and an attractive appearance to an old, badly neglected floor, a solvent-based pigmented seal may be necessary. In this event special seals, based on polyurethane or synthetic rubber, can be used provided certain precautions are taken.

New asphalt floors should be scrubbed with an alkaline detergent solution to remove the surface film, as described above. Where a polyurethane seal is to be used, either clear or pigmented, or a synthetic rubber seal, a test area should first be treated to ensure that the floor is properly prepared and that the seal will adhere to it. A test area is extremely important, as it has already been mentioned that over fifty types of asphalt floors are available, ranging widely in hardness and porosity. A test area will, therefore, ascertain whether the seal will soften and ruin the floor or penetrate sufficiently to obtain satisfactory adhesion, thereby ensuring durability.

The test should be carried out in one of the main traffic lanes, preferably on an area of about 4 m^2 (4 yd^2). A small test area, perhaps alongside a wall away from traffic, is of no value whatsoever and may even give misleading results. The test should be carried out over a period of about four weeks, to allow the seal to harden properly and to be subjected to traffic. If the seal adheres properly after this period and there is no sign of wear or lack of adhesion the floor is suitable for sealing. If, however, the seal blisters or shows any signs of lack of adhesion or wear, then sealing with a solvent-based seal should not be attempted and a water-based seal should be used instead.

Old asphalt floors, previously treated with a water emulsion floor wax, should be scrubbed with an alkaline detergent in water to remove all traces of the floor wax. In the unlikely event of a solvent-based wax having been used it should also be removed with the detergent. Several scrubbings, using metal fibre or nylon web pads under an electric polishing/scrubbing machine may be required to remove all traces of wax. Solvent-based detergent wax removers must not be used as the solvent component will soften and perhaps damage the asphalt.

After cleaning the floor thoroughly, it should be well rinsed and allowed to dry. A test area of seal should then be applied to ensure that the asphalt is suitable for sealing.

If the floor has previously been sealed with a pigmented polyurethane or synthetic rubber seal, the surface should be roughened up with coarse grade metal fibre or nylon web pads under an electric polishing/scrubbing machine and washed to remove all dirt. It should then be mopped over with an approved solvent, and a fresh coat of seal applied immediately after the solvent has evaporated.

It must, however, be stressed that if it is considered necessary to seal an asphalt floor, a water-based seal should be used. Only in special circumstances should a solvent-based seal be applied, and even then only after prolonged tests on site have proved satisfactory.

Waxing Asphalt Floors

A floor wax is an excellent maintenance material for these floors, but discrimination in selection is essential. Solvent-based wax, either paste or liquid, must not be used as the solvent will act on the asphalt and soften the surface.

ASPHALT GROUP OF FLOORS

Water emulsion floor waxes will provide an attractive appearance and facilitate subsequent maintenance.

New floors should be scrubbed with an alkaline detergent solution to remove any surface film. The floor should then be well rinsed with clean water containing a little vinegar or neutralising solution and allowed to dry.

Two coats of water emulsion floor wax should then be applied, preferably coloured to match that of the floor. Maintenance should continue with water emulsion floor wax, applying further coats as required.

Old asphalt floors should be prepared in the same way using an alkaline detergent solution. Repeated applications of a coloured water emulsion floor wax will gradually restore lost colour to a faded floor and greatly improve the appearance.

Use of Detergents

Asphalt floors should be maintained by dry sweeping to remove all loose dirt and soilage, followed by mopping with a neutral detergent solution in water. The floor should then be rinsed and allowed to dry.

Very dirty floors can be cleaned with an alkaline detergent in water and scrubbed with a floor polishing/scrubbing machine and metal fibre or nylon web pads.

It is essential that all oils, fats and greases be removed from asphalt floors at the earliest possible opportunity. Should any have accumulated, detergent crystals should be used to clean the floor. Either the hot or cold water method can be used, as described in the section dealing with concrete and granolithic floors on page 45.

Very strongly alkaline materials such as caustic soda should not be used on asphalt. Coarse paste cleaners and harsh scouring powders should also be avoided. Solvent-based cleaners and detergents must not be used on asphalt.

If a hose is used in the rinsing or cleaning process, the use of extremely hot or cold water should be avoided.

5

THE RESILIENT FLOORS

Included under this heading are linoleum, cork carpet, thermoplastic tiles, PVC (vinyl) asbestos tiles, flexible PVC and rubber. All are floor coverings laid on a sub-floor.

Linoleum was one of the earliest types of resilient floors to be developed and is considered first. Cork carpet, despite its name, is essentially a type of linoleum and its characteristics and methods of maintenance are much more akin to those of linoleum than cork.

Thermoplastic tiles follow next, then a section in which PVC (vinyl) asbestos and flexible PVC tiles, are considered together. The demand for these has grown enormously in recent years.

Development work on both types of floor is currently in progress and improvements and new products are constantly being introduced.

The chapter is concluded with a section on rubber floors.

Cork is often considered to be a resilient floor, but for the reasons given on page 13 it has been included in Chapter 2, dealing with the wood group of floors.

LINOLEUM

The word linoleum is generally used to describe any floor covering having a base of oxidised or polymerised oil adhering to a jute canvas foundation.

Also included in this category is a range of floor coverings in which the hessian is replaced by a bituminised paper felt. Originally introduced because of a shortage of hessian, it is still produced today because of its lower price. Another type consists of linoleum on a felt base, where the pattern is printed on the surface. Felt-backed linoleum has very limited durability and is used almost entirely for domestic purposes.

When oil paint is exposed to the air a rubber-like skin is generally formed on the surface, and it is to this phenomenon that linoleum owes its origin. A British chemist, Frederick Walton, first realised its potential and after a considerable amount of research patented his manufacturing process in 1861. He also devised the word linoleum from two of the main ingredients, linum (flax) and oleum (oil).

Linoleum consists, essentially, of oil, rosin, cork or wood flour, pigments and mineral fillers.

The oil is usually linseed, to which driers are added to assist oxidation and hardening. Rosin is a natural product obtained from trees and to this is generally added resins, which may be natural such as Kauri gum, or synthetic.

Cork flour ranges from the very finest grade used for top quality linoleum to coarse particles of pinhead size in the manufacture of a grade known as cork carpet which will be discussed later. Wood flour is often used instead of cork flour, or as a part replacement.

Pigments are added to give colour and mineral fillers to give body and substance to the mix as well as to reduce the cost.

Linoleum can be manufactured to give many different patterns and a wide variety of types, including plain, jaspé, moiré, marble and others, in both sheet and tile form.

Plain linoleum has a single even colour extending throughout the thickness from the backing material to the surface. Jaspé is made so that different colours are elongated into streaks. Moiré resembles jaspé except that the coloured streaks are feathered into a different formation. In marble linoleum the composition is non-directional and gives an effect resembling marble.

Other types include inlaid, in which the decorative pattern extends through the linoleum to the backing, and embossed linoleum, made either by inlaying or by another process, which puts the design on the linoleum base. There is also printed linoleum, made by stamping the design on the surface. Printed linoleum often resembles embossed linoleum in appearance, but is not suitable for heavy wear as the design tends to wear off when abraded by foot traffic.

Linoleum is made in various thicknesses ranging from about 1·70 mm ($\frac{63}{1000}$ in) to 6·70 mm ($\frac{1}{4}$ in). Linoleum sheet is generally made in widths of 1800 mm (72 in) and tiles are normally 225 mm (9 in) or 300 mm (12 in) square.

Linoleum is used widely in many different locations, ranging from private houses and small offices, where the thinner gauges are laid, to public buildings, theatres, hospitals, ships and hotels, where the thicker gauges are more suitable.

The sub-floor can be of concrete, wood, or many other surfaces. If the floor is subject to rising damp, a damp-proof course should be laid between it and the linoleum.

Adhesives of various types are used. After the linoleum has been laid it is normally pressed with a metal roller of about 68 kg (150 lb) weight to establish a firm bond to the floor. Surplus adhesive should be removed from the linoleum surface as soon as possible after rolling.

Characteristics

Linoleum is made in a wide variety of colours and provides rich scope for the creation of new patterns and designs. Linoleum is also used in special cut-out motifs, many strikingly original and extremely attractive.

This floor covering is resilient rather than hard and will consequently show a deeper indentation than harder surfaces when subjected to sudden impact. Unlike harder surfaces, however, linoleum quickly recovers its original state when the impacting force is removed.

It is a good insulator of sound and is quiet to the tread; it is warm and not tiring to walk or stand on. Linoleum has extremely good durability and if the correct grade is laid and properly maintained it will last for many years. Instances are known where linoleum has given good service for more than 50 years. While it will not crack if laid correctly, linoleum may be cut if subjected to a sharp edge or point load.

The chemical resistance of linoleum is limited. It will withstand only very weak acids for a short time, and is attacked by strong alkalis. Oil and grease may damage linoleum and should not be allowed to penetrate into the surface. The more recently developed types of hardened linoleum, however, have additional resistance to chemicals and are somewhat more durable than conventional linoleum.

Linoleum has good slip resistance properties but can become slippery if wrongly maintained by allowing excess wax to build up on the surface.

Resistance to surface burns is low and although linoleum is non-flammable a lighted cigarette dropped on to the floor will mark or even char the surface. These marks can be removed with a nylon web pad or the finest grade of metal fibre floor pad.

Maintenance

The presence of oil and resin in linoleum limits the materials that can be used for maintenance purposes. Excess water and the use of highly alkaline detergents should be avoided. If too much water is used, it might seep between tiles and loosen them if the adhesive is sensitive to water. It might also percolate to the backing, and cause it to rot or shrink. Continual washing of unwaxed linoleum removes the oil constituent, leaving the surface bare thus allowing dirt to penetrate into the floor. It has been said that more linoleum is ruined by constant washing than by neglect, and so water should be used sparingly.

Strong alkaline detergents, caustic soda and harsh scouring cleaners will literally form a soap with the oils in the linoleum, so removing the oils and causing the linoleum to become hard and brittle, and liable to crack or tear after a short period of time. Strong alkaline detergents may also have an adverse effect on the colour of linoleum. A solution of caustic soda used to clean linoleum has been known to change the colour of a green linoleum to brown.

Newly-laid linoleum should be allowed to settle for one week to allow the adhesive to harden thoroughly. It should then be swept to remove all loose dirt and soilage. This should be followed by a very light scrubbing with a solution of neutral detergent in water, using only the minimum amount of liquid. A light scrubbing will remove any dirt from the surface and prepare the floor for an application of wax. This step is important for new linoleum, as the manufacturer usually coats the surface with a finish which may not be compatible with the floor wax to be used subsequently. The floor should then be allowed to dry and a floor wax applied. Daily maintenance should consist of sweeping followed by buffing, or damp mopping if the floor becomes dirty.

Sealing Linoleum Floors

Sealing linoleum has long been a controversial subject and still is, to some extent. Before water-based acrylic floor seals became readily available various solvent-based seals were sometimes used. The latter, however, all suffer from the defect that no matter how light in colour they are initially, yellowing to some extent takes place on ageing. While this may not be noticeable on some floors, it can be very obvious on others. Turquoise blue, for example, when covered with a pale yellow film assumes a greenish hue, often to a marked degree. Oleo-resinous seals in particular are liable to cause a change of colour.

Solvent-based seals sometimes chip and flake off the surface of the linoleum after a very short period. This is generally due to a lack of adhesion between the seal and linoleum, caused by the presence of the finish applied by the manufacturer. While the poor durability in these instances is not the fault of the seal, remedial action is often lengthy and costly. Touching-up worn seal on linoleum is not easy, particularly if the results are to be examined critically. Removal is also difficult as sanding will inevitably wear away some of the linoleum as well as the seal.

On balance, therefore, sealing with a solvent-based seal of any type is not recommended; particularly as water-based seals are now available which overcome all the defects liable to be found where solvent-based seals are applied.

If the linoleum is old, porous or worn, two thin coats of a water-based seal of the acrylic type should be applied, allowing each coat to dry thoroughly. It is important that the floor is thoroughly clean and dry before any seal is applied. Sealing will fill the open pores of the linoleum and provide a surface for maintenance with a water emulsion floor wax.

The use of a water-based seal can be particularly beneficial on the thicker, or 'battleship' grades of linoleum. Old linoleum of this type frequently requires up to four coats of a conventional water emulsion floor wax to produce a satisfactory finish, because of its extremely porous nature. Two coats of a water-based seal applied initially will enable the required standard to be obtained with one, or at the most two, coats of floor wax. When the floor wax is periodically stripped from the floor with an alkaline detergent, the water-based seal remains unaffected.

If, however, it is required to remove the water-based seal, this can be carried out quite effectively with a fortified alkaline detergent solution in water, together with either a nylon web pad or an abrasive nylon mesh disc, 120 grit, underneath a floor polishing/scrubbing machine. Coarser grits should not be used as they might leave scour marks on the surface of the linoleum.

After removing the water-based seal the floor should be well rinsed and allowed to dry. Two coats of water-based seal should then be reapplied, followed by maintenance with a water emulsion floor wax.

Linoleum which has been treated with an oleo-resinous or other type of solvent-based seal and has not reached the required standard of appearance — perhaps because the seal has worn off in the traffic lanes or become yellow in areas — can be restored to its original attractive state by means of an abrasive mesh disc. 120 grit discs should be used to remove all traces of the solvent-based seal, followed by washing with a neutral detergent solution in water. When dry, two coats of a water-based seal should be applied, routine maintenance being carried out with a water emulsion floor wax.

Waxing Linoleum Floors

Regular maintenance of linoleum with a floor wax is not merely desirable but essential for the best possible service. Floor wax will not only protect the surface from wear by foot traffic but will replace some of the oil and resin that may otherwise become removed by abrasion or repeated washing. This is sometimes called feeding the linoleum, the object being to keep it in a supple and flexible condition, rather than allow it to become hard and brittle.

Linoleum is one of the very few surfaces that can be treated with either a solvent- or water-based floor wax, the choice depending on the circumstances and particularly on the types of floor in adjoining areas. If, for example, neighbouring floors are wood, a solvent-based wax would probably be preferred, so that only one type of floor wax is used throughout the building. If, however, adjacent floors must be maintained

THE RESILIENT FLOORS

with a water-based wax, this type should also be employed on the linoleum. Use of only one type of floor wax will simplify maintenance methods, as well as purchasing, stock control and training procedures.

Perhaps the most important single factor in favour of the use of only one type of floor wax in a building is that of slip resistance. If one kind of wax is used the coefficient of friction between shoe and floor remains almost constant throughout the building and people will adjust themselves to the conditions. If, however, different types of floor wax are used, resistance to slip will vary in different areas, perhaps leading to accidents. If a solvent wax is used on linoleum next to an area treated with a water emulsion floor wax, foot traffic will inevitably carry some solvent wax on to the water emulsion floor wax, causing slippery conditions.

If a solvent-based wax is preferred, either a paste or liquid wax may be used. On some types of thick gauge linoleum it is advisable to apply a paste wax first to provide an immediate finish. Subsequent applications of liquid wax will not then be completely absorbed by the linoleum.

It would be a laborious task to apply paste wax by hand over a large area. Application can be greatly facilitated by dispensing the wax from a heated pneumatic sprayer.

Routine maintenance, particularly on large areas, should be carried out with a liquid wax rather than a paste wax. The former is quick and easy to apply and will clean the floor at the same time as wax is applied. A satisfactory finish can be obtained with either a fine grade metal fibre or nylon web pad under a floor polishing/scrubbing machine, or a dry mop or weighted polishing brush.

Frequency of application will depend on the amount of traffic over the floor, but in general once every two to three weeks will be sufficient. Buffing should be carried out at frequent intervals and preferably daily.

A solvent-based wax should not be used if the linoleum is sealed with a water-based seal, as the solvent may affect the seal adversely.

If a water emulsion floor wax is to be used the floor must be thoroughly clean and dry before wax is applied. Two or even three coats may be required initially, particularly if the floor has not been treated with a water-based seal. The incidence of carbon black heel marks, sometimes a problem on linoleum, can be minimised by the use of a buffable type of water emulsion floor wax, in conjunction with spray or foam cleaning maintenance procedures.

The primary purpose of waxing linoleum is not only to provide correct maintenance so that the floor appears attractive. A good floor wax will fill the pores and make the surface more resistant to moisture, dirt and stains, as well as greatly simplifying routine cleaning. If the floor is sealed with a water-based seal, regular application of a water emulsion floor wax will greatly prolong the life of the seal.

THE RESILIENT FLOORS

Use of Detergents

Day to day soilage can be removed by sweeping or damp mopping with a neutral detergent in water, and the appearance renewed by a light buffing with a floor polishing/scrubbing machine, in conjunction with a fine grade metal fibre or nylon web pad.

Oil and grease should be removed from linoleum as soon as possible with a solvent-based detergent wax remover.

If a build-up of solvent wax becomes evident it can similarly be removed with a solvent-based detergent wax remover. Water emulsion floor wax can be stripped with a solution of an alkaline detergent in water, followed by rinsing. A little vinegar should be added to the rinse water to neutralise any alkali that may remain on the floor. The latter should then be allowed to dry and floor wax re-applied.

It has already been mentioned that strong alkaline detergents should be avoided and this applies also to strong soaps and washing powders. Only a mild detergent should be used for routine maintenance. A neutral detergent is preferred. While a mild soap solution can be used, soap residue could cause slippery conditions to occur over a period of time and for this reason is not recommended. A build-up of soap residue can be removed with a mild abrasive powder and water, in conjunction with a floor polishing/scrubbing machine and medium grade metal fibre or nylon web pads.

CORK CARPET

This is a type of linoleum in which the predominating ingredient is granulated cork, the particle size being approximately equal to that of a pin-head. Other ingredients include oxidised or polymerised oil, colouring pigment and additives similar to those used in conventional linoleum.

Cork carpet is manufactured with a canvas jute backing. For this reason a damp-proof course must be laid underneath the carpet if it is liable to be affected by rising damp.

Thicknesses vary from 2·50 mm ($\frac{3}{32}$ in) to 8·00 mm ($\frac{5}{16}$ in); rather thicker than conventional linoleum. The width is normally 1 800 mm (72 in).

Cork carpet is a luxurious type of flooring used in libraries, nurseries, hotels, hospitals, offices, shops and similar areas. It is also used in sports

THE RESILIENT FLOORS

halls and gymnasia, where its many desirable properties can be utilised to good effect.

Characteristics

The colours of cork carpet are usually rather darker than the normal colours of linoleum. They are restricted in number and are generally similar to those of light, medium and dark cork tile floors.

Cork carpet has properties that are in many ways like those of linoleum, but has a softer, rougher texture. It is also resilient and while it will not withstand heavy point loads, it will recover its original shape to a better extent than cork tiles if ordinary loads are removed from the surface.

Sound insulation properties are excellent and cork carpet is very quiet to the tread. It is warm and has good thermal insulation qualities also.

While the wearing properties are good, it is less durable than linoleum as the surface is rather too soft for very heavy traffic.

Cork carpet is resistant to weak acids and water. The amount of water used for cleaning, however, should be restricted because of the possibility of it seeping between the sheets of cork carpet, and softening the adhesive, or causing the jute backing to rot or shrink. Cork carpet is damaged by alkaline detergents which should be avoided.

Slip resistance properties are excellent, largely because of the rough surface.

Maintenance

Cork carpet has an absorbent nature and if dirt is allowed to enter the surface it can be difficult to remove. Correct maintenance procedures are, therefore, essential if the carpet is to retain its attractive appearance.

The surface of newly-laid cork carpet should be cleaned to remove any trace of surplus adhesive while the latter is still wet. The cork carpet should then be allowed to settle for approximately one week to allow the adhesive to harden thoroughly.

The floor should then be swept to remove all loose dirt and soilage then cleaned, if necessary, with a solution of neutral detergent in the minimum amount of water. Any stains that may remain can be removed with a little solvent-based detergent wax remover and a fine grade metal fibre or nylon web pad.

A solvent-based wax should then be applied and buffed when dry. Routine maintenance should be carried out by buffing, followed by a further application of solvent-based wax as necessary.

THE RESILIENT FLOORS

Sealing Cork Carpet Floors

Because cork carpet is very absorbent, the use of a solvent-based seal has sometimes been advised. This will fill the open pores of the cork carpet and provide a surface for subsequent maintenance.

It should, however, be recognised that the seal will eventually wear and that either touching-up of worn areas or removal of seal will then be necessary. Touching-up could prove difficult and application of a further coat over the whole area may well be preferred. With regard to the removal of seal, a sanding operation would destroy the whole character and appearance of cork carpet, as also would the use of abrasive nylon mesh discs. The only other method of removing the seal is by using a chemical stripping compound, a laborious and difficult task, to be avoided if at all possible.

For these reasons, therefore, while a solvent-based seal could be applied, maintenance with a floor wax only is preferred.

If, however, it is necessary to seal cork carpet, then a solvent-based seal should be used. In this event, it is essential that any finish that may have been applied by the manufacturer be removed before sealing is attempted. If the manufacturer's finish is not removed seal may not penetrate into the cork carpet and subsequently may flake off after a relatively short period.

The manufacturer's finish can be removed with either a mild alkaline detergent solution in water, or a solvent-based detergent wax remover. Tests should first be carried out to ascertain the more effective one before the whole floor is treated. Whichever is chosen, the minimum amount of water should be used and the floor thoroughly rinsed and allowed to dry before any seal is applied. Two, or even three coats of seal may be necessary to provide a smooth, even appearance.

Water-based seals should not be used on cork carpet.

Waxing Cork Carpet Floors

Unsealed cork carpet should be treated first with a paste wax to fill the open pores, then maintained with a solvent-based liquid wax. If, in the course of time, a build-up of wax occurs, it should be removed with a solvent-based detergent wax remover. Maintenance should then continue with a solvent-based liquid wax, buffing at frequent intervals.

Unsealed cork carpet should never be maintained with a water emulsion floor wax, as the water might have a detrimental effect on the flooring material.

While a sealed cork carpet floor can be maintained with a water emulsion floor wax, a solvent wax is preferred. This is because if the

THE RESILIENT FLOORS

seal wears it is better for a solvent wax to come into contact with the cork carpet rather than a water emulsion floor wax.

Use of Detergents

Day to day soilage can be removed by sweeping, or if necessary damp mopping with a solution of a neutral detergent in water, using only the minimum amount of liquid. This should be followed by regular buffing with a floor polishing/scrubbing machine and a fine grade metal fibre or nylon web pad.

When required the floor should be cleaned with a solvent-based liquid wax, which will both remove the dirt and apply a thin coat of wax. This should be followed by buffing, using a fine grade metal fibre or nylon web pad.

A solvent-based detergent wax remover can be used to remove any build-up of solvent wax or spillage of oil or grease. Strong alkaline materials and harsh abrasives should never be used on cork carpet because they will remove the oil constituent and damage the floor. Soap should also be avoided as it tends to leave a deposit which may cause slippery conditions. It might also detract from the appearance and lighten the colour of the floor. Any soap deposit that may be present should be removed with a mild abrasive powder and the minimum amount of water, in conjunction with a floor polishing/scrubbing machine and medium grade metal fibre or nylon web pad.

THERMOPLASTIC TILES

Thermoplastic flooring tiles, sometimes known as asphalt tiles, originated in America and have become widely used in the United Kingdom within the last few decades.

One of the factors contributing to its popularity is that it is among the cheapest of the various types of resilient flooring.

Thermoplastic tiles are manufactured from asphalts or synthetic resins, asbestos fibres, pigments and mineral fillers. There are basically two main types of tile; those based on asphaltic binder and those on plasticised synthetic resins, the latter being available in lighter colours. The raw materials are thoroughly mixed by machine, and the tiles are formed under pressure while hot, and cut to the desired size.

A range of tiles has recently been developed including a proportion of polyvinyl chloride. This provides increased flexibility and enables a wider range of colours to be produced. As the basic characteristics and maintenance procedures are, to all intents and purposes, the same as

THE RESILIENT FLOORS

those for conventional thermoplastic tiles, it is convenient to include them under this heading.

Thermoplastic flooring is available for use in tile form only. The standard size of tile is 225 mm (9 in) square, although other sizes are occasionally made. Thickness is normally 3 mm ($\frac{1}{8}$ in) or 5 mm ($\frac{3}{16}$ in).

These tiles are widely used in commercial and public buildings, hospitals, schools, canteens, shops and in many domestic locations. They must be laid on a level, rigid base, normally a concrete screed with a steel-trowel finish. Thermoplastic tiles are generally heated to make them more flexible and are laid on a suitable adhesive spread evenly on the sub-floor. If, however, the sub-floor is not level, the tiles may crack under the weight of traffic.

While thermoplastic tiles are damp-resisting, damp may rise to the surface along the joints between the tiles. In certain circumstances this could result in a stain being caused by salts from the concrete sub-floor. Prompt cleaning with a damp cloth will usually restore the tiles to their original condition.

Characteristics

The colour range of thermoplastic tiles is not particularly wide and they are normally produced in the darker shades, which may be either plain or marbled. Tiles are of uniform colour throughout their thickness, with the result that the colour shows little, if any, change as the tiles wear.

The colour of some of the darker tiles has a tendency to fade if exposed to direct, bright sunlight. Shades or blinds should be provided, if possible.

Thermoplastic tiles are rather hard and may fracture if subjected to a sudden, sharp impact. They are very sensitive to indentation and are the hardest and least resilient of all the synthetic floor coverings. Tiles are dented by heavy concentrated loads and the legs of furniture should be fitted with furniture shoes to distribute the weight over a wide area.

The tiles are somewhat noisy to walk on. They have good thermal insulation properties and are not cold to the tread. Wearing properties are very good, although they are not intended to withstand heavy industrial trucking.

These tiles have a moderate resistance to most dilute forms of acid and to mild alkaline detergents, although strong alkaline detergents should be avoided. Resistance to water is excellent and floors can be scrubbed repeatedly without harm.

Thermoplastic tiles can be affected by alcohol giving the floor a bleached or white appearance. If alcohol is spilled on a thermoplastic tile floor, only the colour pigments on the surface are liable to be affected, and that area should be scrubbed with a strong solution of neutral detergent and a coarse metal fibre or nylon web pad. This should remove the thin layer of bleached pigments and restore the floor to its original colour. The area should then be well rinsed and allowed to dry.

Oil, grease and solvents readily attack the tiles and soften them. In extreme cases the tiles may disintegrate and need to be replaced. Any spillage should, therefore, be removed immediately. Solvents, for example, white spirit, can also cause the colours to bleed, so that they run into each other and ruin the appearance of the floor.

Thermoplastic tiles have good resistance to slip, although they may become slippery if wet.

Tiles soften when heated and for this reason hot objects should never be placed on thermoplastic tile floors, or the tiles may be permanently disfigured. While thermoplastic tiles are generally satisfactory on floors with underfloor heating, if the surface temperature rises much above $27°C$ ($80°F$) the tiles could be softened.

Radiators should not be directed on to the floor, and heated objects should be prevented with asbestos sheeting or a similar insulating material from coming into contact with the floor.

Very hot water should never be used on thermoplastic tiles, as it might soften them. Only cool or lukewarm water should be used.

Thermoplastic tiles are immune to fungal attack.

Maintenance

Immediately after the tiles are laid any surplus adhesive should be removed from the surface while still wet. They can be walked on as soon as they are laid and this should be encouraged as it assists them to become bedded down. If traffic is not allowed on the floor some tiles may detach from the sub-floor, unless great care has been taken when applying the adhesive to ensure maximum adhesion between tile and sub-floor.

Water should be used very sparingly for the first two weeks after laying, otherwise the adhesive might be softened, resulting in some tiles lifting.

Bare thermoplastic tile is difficult to maintain in a clean and attractive condition. The surface is generally somewhat rough and soil readily adheres to it and penetrates into the open pores of the tile. It is, therefore, important to keep thermoplastic tile floors well waxed with a water emulsion floor wax, not only to retain the attractive appearance of the floor, but also to facilitate subsequent maintenance.

Routine maintenance should consist of brushing, to remove all loose dirt and soilage, followed by damp mopping, if necessary, with a solutic of neutral detergent in water. The use of excess water should be avoided. If required, the floor may be buffed with a fine grade metal fibre or nylon web pad underneath a floor polishing/scrubbing machine. Further coats of floor wax should be applied as required.

Periodically all old floor wax should be stripped from the floor with a solution of mild alkaline detergent in water. The floor should then be well rinsed, adding a little vinegar to the rinse water to neutralise any alkali that might remain, and allowed to dry. Two coats of water emulsion floor wax should then be applied.

Sometimes an old thermoplastic tile floor may assume a whitish or greyish cast. This could be due either to excessive use of strong alkaline detergents affecting the pigments, or to a build-up of water emulsion floor wax. If due to excessive use of detergent, the floor should be scrubbed with a coarse grade metal fibre or nylon web pad underneath a floor polishing/scrubbing machine. Once the whitish cast has been removed the floor should be cleaned and allowed to dry. Two coats of a water-based seal should then be applied, followed by maintenance with a water emulsion floor wax.

If the whitish cast is due to a build-up of water emulsion floor wax, then the floor should be scrubbed as above, but using a solution of alkaline detergent in water. Once the floor is clean it should be well rinsed, adding a little vinegar to the rinse water and allowed to dry. Maintenance should then continue with a water emulsion floor wax.

Oily dust mops and oily sweeping compounds should not be used on thermoplastic tiles as they may cause the tiles to soften. Sweeping compounds should also be free from sand or grit, otherwise the floor may become scratched and marked.

Some sweeping compounds contain a material which will absorb water vapour from the atmosphere. Such a compound should not be allowed to remain for any length of time on the floor, otherwise there is a danger that the floor might become damp and slippery. This can, however, be overcome by sweeping finally with dry sawdust or a dry mop.

Sealing Thermoplastic Tiles

Solvent-based seals must not be used on thermoplastic tile floors, partly because solvents may soften and damage the tiles and partly because of the difficulty of subsequent removal, even if the seal is applied satisfactorily.

Many instances are on record of floors treated in error with a solvent-based seal. The seal has turned yellow over a period of time, with a consequent deterioration in appearance. Removal of seal cannot be carried out either by sanding, or by the use of a chemical stripping compound, as both methods are liable to ruin the floor.

It has been found that perhaps the best remedy is to abrade the seal from the surface with abrasive nylon mesh discs, 100 or 120 grit. Coarser grades should not be used as they may leave scour marks on the surface of the tiles. If the floor is uneven, as is often the case with old thermoplastic tile floors, it may be advantageous to place a metal fibre pad between the disc and drive plate of a single brush floor polishing/scrubbing machine. This will provide additional flexibility and enable the disc to follow better the uneven contours of the floor.

Once all old seal is stripped from the floor it should be vacuumed to remove all dust, and two coats of a water-based seal of the acrylic type applied. A water-based seal will fill the open pores and provide an excellent surface for subsequent maintenance with a water emulsion floor wax.

If, therefore, a seal is necessary on thermoplastic tiles, perhaps because they are worn, pitted or porous, a water-based seal can be used with advantage. Two coats are generally recommended, followed with a water emulsion floor wax.

Waxing Thermoplastic Tiles

Solvent-based waxes should never be used on thermoplastic tiles because of the possible detrimental effects of the solvent. The tiles treated with such a wax, may not show any immediate signs of deterioration, but over a period they could slowly disintegrate.

New thermoplastic tiles are sometimes finished by the manufacturer with a surface coating which prevents the even flow of a water emulsion floor wax, resulting in a spotty appearance. After the initial drying period of two weeks, new tiles should be cleaned thoroughly with a neutral detergent solution in water before floor wax is applied.

Two coats of a water emulsion floor wax should then be applied and the floor buffed, as required, with a floor polishing/scrubbing machine.

If a buffable floor wax is used scuff marks can easily be removed by dry buffing with a fine grade metal fibre or nylon web pad under an electric polishing/scrubbing machine.

Carbon black heel marks often present a problem on thermoplastic tile floors. They can, however, generally be removed with a mild alkaline detergent and a metal fibre or nylon web pad. The incidence of carbon black heel marks can often be greatly reduced, or even eliminated, by the use of a buffable type of water emulsion floor wax, in conjunction with spray or foam cleaning maintenance procedures.

Use of Detergents

While natural and mild alkaline detergents can be used satisfactorily on thermoplastic tiles, scouring powders, paste cleaners and all materials containing solvent should be avoided. Soap should also be avoided as residue may cause slip.

Strong alkaline detergents may cause the tiles to crack or curl, or the colours to bleed. If damage has taken place, the tiles should be treated with an abrasive nylon mesh disc, 120 grit, followed by vacuuming to remove all dust. Two coats of a water-based seal should then be applied, followed by a water emulsion floor wax.

Thermoplastic tiles can vary widely in quality, and unless poor tiles are laid with great care the edges may not fit flush. Gaps may appear between the tiles and if the floor is cleaned with an alkaline powder detergent a build-up of powder may take place in the gaps. If a water emulsion floor wax is then applied to the floor, white lines, up to about 13 mm ($\frac{1}{2}$ in) wide, may appear along the lines of the gaps. This is caused by the action of the alkaline detergent on the water emulsion floor wax, breaking the emulsion down to a fine powder.

Perhaps the best remedy is first to remove the emulsion floor wax with a liquid alkaline detergent in water, in conjunction with a floor polishing/scrubbing machine fitted with a scrubbing brush. The latter is preferred to a floor pad because the bristles are better able to search into the gaps between the tiles and loosen the accumulated dirt and detergent.

The floor should then be well rinsed, adding a little vinegar to the rinse water, and allowed to dry. Two or three coats of a water-based seal should then be applied. This will prevent any detergent remaining in the gaps from attacking subsequent applications of water emulsion floor wax used for maintenance.

When properly installed and maintained, thermoplastic tiles, although generally cheaper, are likely to give as long and satisfactory service as any other type of resilient floor covering.

PVC (VINYL) ASBESTOS AND FLEXIBLE PVC

There are two main types of polyvinyl chloride (PVC) floor coverings in use today; PVC (vinyl) asbestos and flexible PVC, the latter sometimes known as pure vinyl. As the maintenance materials and methods used on the two types of floor covering are very similar, it is convenient to consider them together.

Although the first PVC floor was laid in the early 1930s, it is only within the last two decades that PVC (vinyl) asbestos has increased in popularity to the extent that it now appears to dominate all other types of resilient floor covering.

PVC (vinyl) asbestos consists of a blend of polyvinyl chloride polymer or copolymers, asbestos fibre, fillers and pigments, to which are added suitable plasticisers and stabilisers. Flexible PVC consists essentially of the same raw materials, but without asbestos fibre and is often finished with a very smooth, glossy surface.

It should be recognised that while the above are the two main types of PVC floor covering, tiles with intermediate characteristics are produced by varying the relative amounts of polyvinyl chloride and asbestos fibre, the latter included as a reinforcing agent and to reduce the raw material cost. It is frequently very difficult to distinguish between a true flexible PVC tile and one containing a small proportion of asbestos fibre, technically a PVC (vinyl) asbestos tile.

PVC (vinyl) asbestos is produced in tile form only, the most common sizes being 225 mm (9 in) square and 300 mm (12 in) square and ranging in thickness from 1·6 mm (0·063 in) to 3·2 mm (0·126 in).

Tiles are normally plain or mottled and both types have the colour distributed throughout the thickness of the tile. Tiles having special designs penetrating only a fraction of the thickness are sometimes manufactured for particular requirements, but these are generally not suitable for very heavy traffic areas.

Flexible PVC is manufactured in both sheet and tile form. Sheet flexible PVC is generally supplied in widths ranging from 900 mm (3 ft) to 1 800 mm (6 ft), and tiles are normally 300 mm (12 in) square. Thicknesses range from 1·57 mm (0·062 in) to 3·18 mm (0·125 in) for both sheet and tile forms.

Flexible PVC may be unbacked or backed with a variety of materials, and new backings are still being introduced. Hessian is one of the most common backing materials and others include felt, plastic foam, cork or fabric, the latter being added after the flexible PVC sheet has been calendered.

PVC (vinyl) asbestos tiles can be laid on a variety of sub-floors, including concrete and asphalt. A damp-proof course is unnecessary, provided that there is no danger of a considerable amount of moisture rising through the sub-floor. The tiles are suitable for underfloor heating if the surface temperature does not exceed $27°C$ ($80°F$) and if they are laid with a suitable adhesive. They must be laid warm and at the moment when they are laid the tiles should be pliable, so that they may follow exactly the contours of the sub-floor.

Flexible PVC can also be laid on a variety of sub-floors, except that an adequate damp-proof course is necessary on concrete if there is any possibility of rising damp. They are suitable for underfloor heating, up to a surface temperature of $29°C$ ($85°F$). Both tiles and sheet are stuck

THE RESILIENT FLOORS

with a suitable adhesive to the sub-floor. Sheet flexible PVC can also be welded at the seams to produce a jointless floor and this can be particularly advantageous in areas where excessive spillage is liable to occur.

PVC (vinyl) asbestos tiles are widely used in commercial buildings, shops, schools, hospitals, public buildings, and a great variety of both industrial and domestic situations.

Flexible PVC is somewhat more expensive and is found in hospitals, commercial buildings, schools, public buildings, shops and also in domestic locations. Backed types of flexible PVC, particularly those with plastic foam or cork, are luxurious and quiet and are particularly suitable for hospitals, libraries, corridors and offices where quietness is desirable or essential. Welded sheet flexible PVC is used extensively in hospitals for hygienic reasons, as water and dirt cannot lodge in cracks between sheet or tiles.

Characteristics

PVC (vinyl) asbestos tiles are manufactured in a wide range of bright, clear and attractive colours, from white through pale and dark shades to black. Pleasing patterns can be made and a large selection of colour schemes designed.

Flexible PVC is also produced with a wide choice of colour and pattern. Some types of flexible PVC are produced in a single main colour, incorporating coloured plastic chips, which can give an inlaid effect.

PVC (vinyl) asbestos tiles are moderately hard with more flexibility than thermoplastic tiles, but very much less than flexible PVC. PVC (vinyl) asbestos tiles bend only slightly before they break, but flexible PVC in both sheet and tile form can be rolled or bent without breaking. PVC (vinyl) asbestos tiles have only moderate resistance to indentation and the maximum weight that the flooring can sustain without a permanent indentation is relatively light. While the tiles are more resilient than thermoplastic tiles, they are less resilient than flexible PVC, linoleum and rubber. The legs of furniture and other heavy objects should be fitted with furniture shoes to spread the weight over a wider area.

Most flexible PVC is not quite as resilient as rubber, but plastic foam and cork backed varieties have good resilience and recovery properties.

PVC (vinyl) asbestos tiles are somewhat noisy to the tread, but are, perhaps, a little quieter than thermoplastic tiles. Flexible PVC varies widely from a noise level similar to that given by PVC (vinyl) asbestos tiles to extreme quiet. Plastic foam and cork backed flexible PVC are among the quietest of all types of resilient floor covering and equal cork tiles in this respect.

THE RESILIENT FLOORS

Both PVC (vinyl) asbestos and flexible PVC floorings are not cold to the tread. The various types of backed flexible PVC have very good thermal insulation properties and are warm to walk on.

The wearing properties of both types of PVC are very good under normal foot traffic conditions. They are not intended to withstand heavy industrial wear.

PVC (vinyl) asbestos tiles have good resistance to mild acids and most common chemicals. They will also resist alkaline detergents, and have an excellent resistance to water. Some grades resist oil and grease extremely well and these are suitable for use in kitchens and similar areas. They are, however, softened by solvents, for example white spirit, particularly if allowed to remain on the surface for some time. Solvents may also cause colours to 'bleed', so that they run into each other and spoil the appearance of the floor.

Flexible PVC tiles and sheet have very good resistance to weak acids, alkalis and most common chemicals. They also have a high resistance to oil and grease and to water. Top quality flexible PVC is also resistant to white spirit and similar solvents, but some of the lower quality grades are adversely affected. It is, therefore, advisable not to allow solvents to come into contact with flexible PVC.

Both types of flooring normally have good resistance to slip, although they may become slippery if wet.

Both PVC (vinyl) asbestos and flexible PVC are liable to be marked by cigarette burns, sometimes to a considerable extent. Normal techniques for the removal of burn marks, for example using steel wool and an abrasive paste, may prove only partially successful if the marks are extensive. At best, removal will result in a slight depression as the floor surface is physically abraded away. Prevention is better than cure and plenty of ashtrays should always be provided if there is any possibility of damage by cigarette burns.

Both PVC (vinyl) asbestos and flexible PVC are immune to fungal attack.

Maintenance

As soon as the floor is laid any surplus adhesive should be removed from the surface while still wet. PVC (vinyl) asbestos tiles can be walked on immediately after laying and foot traffic will assist them to become bedded down. Traffic can also be allowed on flexible PVC floors as soon as they are laid.

Floors should not be washed for seven days after laying to allow the adhesive to harden. During this period cleaning should be carried out by brushing. For the next seven days water should be used sparingly

THE RESILIENT FLOORS

and if cleaning other than brushing is necessary, it should be carried out with a mop, wrung out until almost dry.

Once the floor is properly bedded down, the best protection against abrasion by foot traffic and carbon black heel marks is given by a water emulsion floor wax. The floor should be scrubbed with a neutral detergent in water, rinsed and allowed to dry. Two coats of a water emulsion floor wax should then be applied.

Routine maintenance should consist of brushing, followed by damp mopping, if necessary, with a solution of neutral detergent in water. If required, the floor should be buffed with a fine grade metal fibre or nylon web pad underneath a floor-polishing/scrubbing machine. Further coats of floor wax should be applied as required.

Periodically all old coats of emulsion floor wax should be stripped with a solution of mild alkaline detergent in water. The floor should then be well rinsed, adding a little vinegar to the rinse water and allowed to dry. Two coats of water emulsion floor wax should then be reapplied.

Oily dust mops and oily sweeping compounds should not be used on PVC floors, as they might cause PVC (vinyl) asbestos tiles to soften and can cause deterioration and slippery conditions on flexible PVC floors.

Sealing PVC (vinyl) Asbestos and Flexible PVC Floors

Conventional solvent-based seals must not be used on these floors because of the possible detrimental effects of the solvent.

If, however, a PVC (vinyl) asbestos floor has been treated with a solvent-based seal and removal of the seal is required, it can be abraded from the surface with abrasive nylon mesh discs, 120 grit. as described in the section dealing with thermoplastic tiles on page 81. The use of discs on flexible PVC is not recommended as the floor surface is sometimes very thin and permanent damage may result.

Once all old seal is removed from the PVC (vinyl) asbestos floor, two coats of a water-based seal should be applied, followed by maintenance with a water emulsion floor wax.

Waxing PVC (vinyl) Asbestos and Flexible PVC Floors

Floor waxes containing solvent should not be used because of their possible detrimental effects. While it has always been recognised that PVC (vinyl) asbestos tiles should be maintained with a water emulsion floor wax, some manufacturers of flexible PVC flooring claim that only sweeping and mopping with clear water are needed to maintain their floors. Those who do not recommend waxing or the use of detergents imply that their use merely incurs an unnecessary expense.

A survey, however, among the leading manufacturers of flexible PVC floors, shows that the vast majority are favourably inclined towards waxing. It is recognised that in time wear and tear must inevitably cause a deterioration of the flooring surface. Application of a water emulsion floor wax improves the appearance, protects the flooring surface from the abrasive effects of foot traffic, minimises scratching and facilitates subsequent maintenance. Slip resistance characteristics can also be improved by discrimination in selection of the floor wax. Not only do flexible PVC floors look more attractive when waxed, but they also give better and longer service.

After the initial clean, two coats of a water emulsion floor wax should be applied and the floor buffed, as required, with a floor polishing/scrubbing machine. On new PVC (vinyl) asbestos tiles a third coat may be necessary to provide an even appearance and adequate protection, because of the porosity of the tiles. A third coat will seldom be necessary on flexible PVC because the surface is generally smoother and less porous.

If a buffable material is used, scuff marks can easily be removed by dry buffing with a fine grade metal fibre or nylon web under an electric polishing/scrubbing machine.

Carbon black heel marks sometimes present a problem, but they can generally be removed satisfactorily and subsequently prevented by the use of spray or foam clean maintenance procedures.

When waxing PVC (vinyl) asbestos tiles it is important that a proper, continuous film of water emulsion floor wax be maintained on the surface. If too little is applied, foot traffic will wear through the protective layers of floor wax and dirt will penetrate into the tiles themselves, instead of being retained in the wax. The appearance of the floor will rapidly deteriorate, carbon black heel marks may become prominent and the surface will lose gloss.

If too much floor wax is applied, powdering, yellowing and scratching could result. In this event, it is recommended that all wax be stripped from the surface and the floor retreated with two coats of a buffable water emulsion floor wax.

A slightly different procedure should be adopted when waxing flexible PVC floors, because of their relatively non-porous surface. For this reason water emulsion floor waxes should be applied sparingly to provide a thin, even coat. If mops are used they should be wrung out until they are only just damp to ensure that the bare minimum amount of floor wax is applied. The wax must not, however, be spread too thinly, or the result will be a patchy appearance with low gloss. Two coats are normally required, the second coat being applied only after the first is hard dry, otherwise low gloss and a streaky appearance may result.

Water emulsion floor waxes frequently take longer to dry on flexible PVC than on other floors. This is because this type of PVC is less porous, therefore more liquid remains on the surface and as drying takes place by evaporation of water, the process is slower.

Use of Detergents

Neutral, mild and even strong alkaline detergents can be used satisfactorily on PVC floors. It is always advisable, however, to use the weakest material necessary to achieve the desired results, as strong solutions may seep down between the tiles and affect certain adhesives.

Scouring powders and paste cleaners can be used on PVC (vinyl) asbestos floors, if necessary, but they should not be used on flexible PVC. Even the mildest of abrasive powders may cause scratch marks on smooth flexible PVC floors. Similarly, when a flexible PVC floor is scrubbed by a machine, only the finest grade of pad should be used to minimise the possibility of scratch marks on the surface.

Solvent-based detergents should not be used on PVC as solvent may damage the floors. Soap also is not recommended as soapy residues may cause slippery conditions.

RUBBER

Rubber floors first came into prominence at the beginning of this century and have since proved very successful. Perhaps the qualities that have contributed most to their success are durability, luxurious appearance and quietness to the tread.

While the term 'rubber' originally referred to material of a natural origin, developments in recent years in both the raw material and manufacturing areas, have resulted in the production of a variety of synthetic materials resembling rubber, to satisfy a variety of requirements.

The basic materials used in the manufacture of rubber for floors are natural or synthetic rubber, or a mixture of both, together with other compounds. These include mineral fillers such as china clay, pigments which are insoluble in water and which provide the colours, and additives such as anti-oxidants.

It is generally considered even today that the best quality rubber for floors contains a high proportion of natural, rather than synthetic, rubber. While under normal conditions a flooring material consisting primarily of natural rubber is extremely suitable, special circumstances may require a type of rubber based on synthetic materials. Synthetic has advantages over natural rubber with regard to resistance to oils, greases and solvents and is sometimes specified accordingly.

THE RESILIENT FLOORS

Rubber for floors is produced in a number of different types, either in sheet or tile form.

Sheet rubber is generally made in 900 mm (3 ft) to 1 350 mm (4 ft 6 in) widths, although widths up to 1 800 mm (6 ft) are sometimes produced. The normal thickness varies from 3 mm ($\frac{1}{8}$ in) to 13 mm ($\frac{1}{2}$ in), the most common for normal commercial floors being 5 mm ($\frac{3}{16}$ in).

Tiles are cut from sheet rubber or moulded separately. They usually range from 225 mm (9 in) to 450 mm (18 in) squares, although other sizes are also made. The edges may be square cut or interlocking. Sheet and tile rubber are widely used in hotels, hospitals, libraries and numerous other locations.

Grease-resistant rubber flooring is made from synthetic rubber and is designed to withstand spillage of oil, grease and similar materials, and is used, for example, in canteens, kitchens and ships.

Sponge rubber flooring consists of a solid rubber walking surface with a sponge rubber or cellular foam backing, sometimes with a fabric between the two layers. It was first made about forty years ago, but it is only in comparatively recent times that its main qualities of softness and quietness have been fully appreciated. It is among the most silent of all hard surface floor coverings and is widely used in hotels, cinemas, hospitals, nurseries and libraries, where quietness is required. Thicknesses vary, but are similar to those of sheet rubber floors.

Ribbed or ridged tiles comprise thick rubber compounds, often with ribs also on the underneath which can bind to a wet concrete base. As the tiles are firmly secured to the sub-floor they will withstand heavy traffic, including trolleys and trucks. They are used in such places as shopping precincts, air terminals and railway stations. Lighter gauge ribbed tiles are often used to provide a non-slip surface for the surrounds of swimming baths and in changing rooms. Studded tiles are also produced to give a slip resistant surface.

Rubber flooring can be laid over almost any level surface that is smooth and in good repair. If covering a new wood board floor it is advisable first to lay plywood or hardboard over the wood, as there is a possibility of warping of the wood boards which may show through the rubber. Concrete should be thoroughly dry before rubber is laid.

Most rubber floors are secured with an adhesive, suitable for the sub-floor on which the rubber is laid.

Characteristics

Sheet and tile rubber are produced in a wide variety of bright, attractive colours and in marbled patterns. Pleasing designs can be produced and rubber floors generally have an aristocratic appearance.

THE RESILIENT FLOORS

Ribbed or ridged tiles used for heavy traffic and industrial areas are generally made in black only.

Rubber is extremely resilient and soft to the tread. Added softness is obtained from those types backed with sponge rubber or cellular foam.

It is always advisable to protect rubber floors from indentation by heavy objects. Flat furniture shoes should be placed under point loads, for example small castors, to distribute the weight over a wider area.

Rubber is extremely silent underfoot, one of its main attributes. It is also warm and has good thermal insulation properties. It is softened by heat, however, and hot objects should never be placed on a rubber floor. It is also not suitable for use where there is underfloor heating.

Wearing properties of rubber are extremely good. Under normal traffic conditions a rubber floor should give excellent service for at least 25 years if laid and maintained correctly. Some rubber floors are still in good condition after 50 years, although this is exceptional.

The chemical resistance properties of rubber vary widely depending on the proportion of natural to synthetic rubber in the material. A flooring material consisting essentially of natural rubber, for example most sheet and tile rubber floors found in commercial buildings, has a low resistance to oil and grease and will be softened by white spirit, petrol and similar solvents. Solvents can also cause localised swelling and stretching. The result is that after a period the rubber may lift from the sub-floor, causing a ripple effect. The only remedy in this event is to cut out the affected areas and relay new rubber. Solvents can also cause bleeding of colour, so that colours dissolve and run into each other.

Rubber has very good resistance to water and can be washed repeatedly without harm. It is also resistant to dilute acids, but is affected by strong alkaline detergents.

Rubber flooring based mainly on synthetic rubber has a similar resistance to water, acids, and alkalis, but a much improved resistance to oils, greases and solvents. The floors have excellent resistance to slip even when highly waxed.

A characteristic of rubber floors is their excellent electrical insulation properties. Special anti-static rubber floors are discussed on page 126 in Chapter 7.

Rubber floors are particularly hygienic and resistant to mould growth and bacterial decay. They are also immune from pest attack.

Maintenance

Following installation, any surplus adhesive should be removed from the surface of the rubber while it is still wet. The floor should not be

THE RESILIENT FLOORS

washed for seven days after laying to allow the adhesive to harden thoroughly.

Possibly the main factors that cause deterioration of these floors are oxygen and sunlight, which accelerate the oxidation process. If not protected, rubber will slowly oxidise and become rough on the surface, which will eventually lead to cracking. This can be prevented by maintenance with a water emulsion floor wax.

If, however, the surface of a rubber floor has become rough, it can be remedied and the floor restored to its original attractive condition by the use of abrasive nylon mesh discs. The floor should be cleaned and treated with an abrasive nylon mesh disc, 100 or 120 grit, under a floor polishing/scrubbing machine. A coarser grade disc should not be used as scouring marks may remain on the surface. Alternatively, if the surface of the rubber is only very slightly oxidised, treatment with a metal fibre or nylon web pad, medium grade, may give the required results. Two or three buffings may be required, finishing with a fine grade pad to produce a smooth surface.

The floor should then be vacuumed and damp mopped to ensure that all particles of metal fibre or nylon web are removed, followed by treatment with a water-based seal and water emulsion floor wax.

Normal maintenance of rubber floors should consist of brushing to remove all loose dirt and soilage, followed by damp mopping, if necessary, with a solution of a neutral detergent in water. The floor should be maintained with a water emulsion floor wax, preferably of the buffable type and this should be buffed regularly. When buffing a water emulsion floor wax with a floor polishing/scrubbing machine, it is essential to use the finest type of pad necessary to achieve the desired results. It should be recognised that some of the very smooth, glossy rubber surfaces can be scratched by even fine grade pads. The machine should not be allowed to remain in one spot for any longer than is absolutely necessary to ensure that a pad does not wear through the film of floor wax and scratch and dull the surface of the rubber.

Ribbed or ridged heavy duty tiles or matting, normally need only to be brushed to remove all loose dirt. This should be followed by vacuuming, if necessary. Occasional washing with a neutral detergent in water will remove all normal dirt. If, however, the ribbed tiles or matting become very dirty, they can be cleaned with a solution of a mild alkaline detergent in water, scrubbing with a stiff broom. The latter should be dipped into the detergent solution and the floor brushed along the grooves between the ribs until clean. The floor should then be thoroughly rinsed and allowed to dry.

Oil, grease or solvent spillage on a natural rubber floor should be removed immediately to prevent the rubber from being harmed. A cloth can be used to remove small spillages, but if a large area is affected a mild alkaline detergent should be used to emulsify the material and enable it to be easily removed.

Oily sweeping compounds and dust mops must not be used on rubber floors as they may stain and soften the surface.

Sealing Rubber Floors

Solvent-based seals may soften and stain rubber floors. They could also prove extremely difficult to remove if subsequently this became necessary, and, for these reasons, should not be used.

If a floor seal is necessary, for example to re-surface an old rubber floor after it has been cleaned and smoothed with abrasive nylon mesh discs, a water-based seal of the acrylic type should be used. This will fill any open pores and provide an excellent foundation for subsequent maintenance with a water emulsion floor wax.

Waxing Rubber Floors

Solvent-based waxes should never be applied on rubber floors because solvent might soften the rubber. Water-based waxes only should be used.

Before a water emulsion floor wax is applied to rubber the surface must be perfectly clean and this should be achieved with a solution of neutral detergent in water. It is essential that only the minimum amount of water be used as excess might enter the cracks between the rubber tiles or sheets and affect the adhesion of the rubber to the sub-floor. Penetration of water might also cause lifting at the joints.

Once the floor is clean and dry, a water emulsion floor wax should be applied. The use of a floor wax is essential for correct maintenance, particularly immediately after an initial clean. It protects the surface of the rubber from abrasion by foot traffic and assists in the production of a dirt-resisting surface.

A floor wax also prevents oxygen in the atmosphere from coming into contact with the rubber and therefore minimises possible deterioration by air and sunlight.

A top quality water emulsion floor wax with a high wax : polymer ratio is preferred to one with the reverse. This is because wax is more flexible than most polymers and will bend easier with the rubber floor when subjected to movement by traffic. Materials with a high proportion of wax are also easy to buff and will readily respond to spray or foam cleaning maintenance procedures. Floor waxes with a high proportion of polymer may require heavier equipment to achieve the desired results, with possible detrimental effects on the rubber. A floor wax with a high proportion of wax is also easier to remove when periodic wax stripping becomes necessary.

In general, when applying a water emulsion floor wax to rubber a greater coverage can be obtained than on many other surfaces, for example linoleum. This is because rubber has a smoother surface and is relatively non-absorbent. Water emulsion floor waxes sometimes dry slower on rubber, for the same reason. A second coat should never be applied before the first is hard dry, otherwise low gloss and streaking might result.

Use of Detergents

Darkening of rubber floors can occur if water emulsion floor wax is allowed to build up in areas. It will probably be a very gradual process and may not be noticeable. Spot cleaning with a mild abrasive paste cleaner will quickly show whether a build-up of wax has taken place. If wax stripping is necessary, it can be carried out with a solution of a mild alkaline detergent in water, in conjunction with a floor polishing/scrubbing machine and fine or medium grade metal fibre or nylon web pads. When clean, the floor should be thoroughly rinsed, adding a little vinegar to the rinse water to neutralise any alkali that might remain. The floor should then be allowed to dry and a water emulsion floor wax re-applied.

Strong alkaline and caustic detergents may prove harmful to rubber and should not be used. Coarse abrasives, harsh scouring powders, solvent-based cleaning compounds, soap powders containing caustic soda and liquid soap containing oils, such as pine oil, should also be avoided.

The use of any type of soap is not recommended, as some residue will inevitably remain and could cause slippery conditions.

The old adage 'prevention is better than cure' is particularly true concerning rubber, but if correct maintenance procedures are adopted as soon as the floor is laid, rubber will give many years of trouble-free service.

6

CARPET GROUP OF FLOOR COVERINGS

One of the earliest types of carpet was made from hair felt or black wool and dates back to the fourth or fifth century B.C. Pile carpets were in existence in A.D. 3 and handwoven carpets have been used in the East for many hundreds of years. It was in the sixteenth or seventeenth century that carpets were first manufactured in Great Britain. Since this time their use has increased enormously and many new types have been developed.

Carpets have long been associated with luxury. No other floor covering can compare with them for appearance and comfort. They are noted for their warmth and also for their non-slip qualities, a very important factor in some locations.

It is recognised that carpets are relatively easy to maintain, particularly when compared with some types of resilient floor. Perhaps the main reason for their not having been used more extensively in the past was because of their comparatively high initial cost. Developments over recent years, however, have resulted in many less expensive types becoming available, so that this factor no longer applies in many instances.

Despite the higher initial cost of carpet compared with many other types of floor covering, it has been found that, in some cases, maintenance costs are lower, so that a carpeted floor can become more economical over a period of years. Cost analyses have been carried out to determine whether or not the installation and maintenance costs of carpet and other floor coverings equate. There are, of course, many variables in an investigation of this kind, particularly as new types of carpet are constantly being introduced. Methods of maintenance of both carpets and other types of floor covering are constantly being developed and improved all the time, so that maintenance costs can vary widely depending on the circumstances. In very general terms, however, it has been estimated that installation and maintenance costs of carpet and resilient floor coverings, the generally accepted alternative to carpet, are about equal after a period of 15 to 25 years.

For a great many years the presence of carpet indicated a special status, but today it is specified for a very wide variety of locations. The tremendous increase in the use of man-made fibres in carpets that has

CARPET GROUP OF FLOOR COVERINGS

taken place in recent years has made it possible for carpet to be laid in areas where previously it would not have been considered.

In addition to domestic premises, carpets have long been found in hotels, restaurants, clubs, offices and many public buildings. They are now finding an increasing use in a wide variety of commercial buildings, office blocks, hospitals and schools, as well as many other locations.

The demand for carpet is increasing rapidly, particularly with the advent of soil-resistant synthetic fibres. Knowledge of maintenance is, therefore, essential, but the correct materials and methods needed will vary according to the materials used in the manufacture and the method of construction of the carpet concerned.

Before considering maintenance, the materials used in the construction of carpet, together with their characteristics, will be discussed.

The word carpet is defined as 'textile fabric for covering floors' and embraces many different materials. Carpet has long been recognised as being warm and soft to the tread, very different from other types of floor. Today, however, carpet has become harder and other softer forms of resilient floor have been developed and the difference is less marked. Tufted, flocked and needled carpets are very different in appearance and properties from conventional woven carpets and are sometimes termed carpeting to distinguish them from more traditional types.

Carpets are constructed, essentially, from fibres which may be natural or synthetic in origin, and backings of various types. Dyes are used to provide colour and these can vary widely in quality, particularly with regard to fastness to light and liquids.

Carpet Fibres

Natural

Natural fibres include wool, cotton and flax, of which wool is undoubtedly the most important in carpets.

Persian, Chinese and Turkish carpets, produced for many hundreds of years and acknowledged to be of the highest quality, consist of pure wool. Even today, despite the advent of many different types of synthetic fibre, wool is often considered to be the best material for carpet pile.

Wool is unique because it has the greatest resilience of all the fibres which can be used to make carpet pile. Wool carpets tend to retain their appearance better than those manufactured from synthetic materials, as the pile recovers quickly after it has been pressed down.

CARPET GROUP OF FLOOR COVERINGS

Wool can absorb moisture from the atmosphere and normally contains about 15 to 18 per cent moisture, compared with from about 0 to 3 per cent moisture for most synthetic materials. It also retains its attractive appearance longer than most other fibres and does not show soil as readily.

Wool varies in its characteristics according to its country of origin. When used in the manufacture of carpets, different types of wool are selected and blended to give the required properties, such as resilience, warmth and durability.

Cotton fibres are fine in texture but vary in size and can easily be confused with rayon. If not properly refined, cotton may discolour on ageing: it has not the resilience of wool and will not recover as well after it has been pressed down.

Flax, sometimes known as linen, can resemble cotton or rayon when bleached. It is sometimes used for the warp in rugs and may be blended with other fibres. One of its characteristics is that it is likely to suffer discoloration if treated with alkaline materials.

Synthetic

Synthetic fibres are generally more uniform in size and smoother than natural ones. Because of this, they are normally easier to clean. They also absorb water to a lesser extent than do natural fibres, so that drying is quicker.

Many raw materials are used in the manufacture of synthetic fibres, including wood pulp, petroleum and coal. Perhaps the most important synthetic fibres used in carpets consist of cellulose, nylon, polyester, acrylic, polypropylene and elastomer materials.

The oldest type of synthetic fibre is rayon, made from wood pulp cellulose. Special types of rayon with high abrasion resistance are used in carpets. It is less resilient than some other carpet fibres and may soil more easily. Modified rayon fibres have been produced which have greater durability and better resistance to crushing and soiling than standard rayon. Modified rayon is probably the most widely used synthetic fibre in carpets today and new types are constantly being developed.

Good quality carpets are produced by blending rayon with other fibres.

Nylon, also known as polyamide, was the first truly synthetic fibre. It is very strong, has excellent resistance to abrasion, very low moisture absorption and cleans easily. It also blends well with other types of fibre to produce very durable carpets and is frequently included with wool for this purpose.

CARPET GROUP OF FLOOR COVERINGS

Under conditions of low humidity static electricity can build up on nylon carpets. This is discussed further on page 101.

Polyester fibres are also strong, non-absorbent and have good resilience. Developed in comparatively recent years, it is finding increasing use in carpets, particularly in conjunction with other fibres.

Acrylic fibres were first introduced into the United Kingdom carpet market in 1957 and by the early 1960s were firmly established. They have good resilience, resistance to abrasion and stains, and are widely used in carpets, either as the sole material or blended with other fibres such as nylon.

These carpets are normally warm, light in weight, soft to the tread and they resemble wool somewhat in handling. They are also comparatively easy to clean.

Polypropylene fibres are also synthetic, non-absorbent and light in weight. They are characterised by being stronger than nylon. Polypropylene is used to a limited extent as a carpet pile; its main use is in carpet backing.

Elastomeric materials are so called because they possess flexible, rubber-like properties. They are also known in the U.S.A. as spandex materials. Polyurethane synthetic fibres are, perhaps, the most important in this group. Elastomeric materials are found only to a limited extent in carpets.

It has already been mentioned that different types of fibre are frequently blended to improve specific qualities of a carpet. Many Axminster and Wilton carpets are now made of a blend of 80 per cent wool strengthened with 20 per cent nylon. Other blends include wool/rayon, acrylic/rayon and nylon/rayon. Some blends consist of three different fibres, for example wool/rayon/nylon.

Each blend is intended to fulfil specific requirements, for example low cost with good durability.

Carpet Backing

The material used for backing a carpet is extremely important and will have considerable effect upon the life of the carpet. Carpets are backed with materials which may be of natural or synthetic origin. Natural materials include glue, starch and rubber: synthetic materials include resins and rubber or blends of both.

Many carpets with synthetic resin or synthetic rubber backing now have an additional backing of jute, which is intended to lengthen their life.

CARPET GROUP OF FLOOR COVERINGS

Carpet Dye

Dyes used in carpets can be affected by light, oxygen and liquids. Loss or change of colour can result in a noticeable difference in appearance. Some oriental carpets are touched up occasionally by hand with dyes which may be fugitive.

When cleaning a carpet, therefore, thought should be given to possible adverse effects on the dyes. Tests should be carried out by using the shampoo at the correct concentration to ensure that the dyes are fast. This process is discussed in greater detail on page 109.

Classification by Construction

Carpets are frequently classified according to the method of construction, a very important factor in both durability and maintenance considerations. Carpets are constructed by many different methods and examples include Axminster, Wilton, Brussels, tapestry, tufted, needleloom or needle-punched felt, knitted, pile-bonded, electrostatically-flocked, and various types of tile. New kinds of carpet are constantly being developed. Types of pile vary widely from low cut with a tight weave to high pile with a loose weave: In general, as the depth of pile increases, so the frequent need for cleaning increases to give the best possible results.

Axminster carpets can be manufactured in multi-coloured designs and are extremely attractive in appearance. They are very durable and have gained an excellent reputation through the years.

The finest Wilton carpets are characterised by strength and body, as the method of manufacture enables a dense fabric to be achieved with very secure tufts. They are extremely durable and luxurious in use. Like Axminster carpets, they are firmly established as carpets of the highest quality.

Brussels carpets are woven by the same method as that used for Wilton, except that the looped pile is uncut. Brussels carpets are very serviceable and the problem of flattening as happens with poorer quality fibres, is reduced.

Printed tapestry is sometimes used for carpets but has now largely been replaced by other types of carpet.

Tufted carpets were first introduced into the United Kingdom at the end of the 1950s and it has been estimated that at the present time they account for over one half of the total amount of carpet produced. In view of the very rapid expansion in this field, and the fact that they are now widely used in offices and commercial premises, they will be dealt with in rather more detail than will the other types of carpet.

CARPET GROUP OF FLOOR COVERINGS

Tufted carpet is sometimes referred to as contract carpet because of its comparatively low cost and suitability for commercial use. It is manufactured by a high speed process. Yarn is needled into a backing at a very rapid rate and the tufts, which can be either looped-pile or cut, are fixed in place with an adhesive.

The fibres used in its manufacture range from wool to nylon and may include acrylic resins and polypropylene.

Needleloom, or needle-punched felt, consists essentially of synthetic surface fibres, which may be polypropylene, nylon or acrylic resins, compacted by needling from an original loose layer about 75 mm (3 in) thick, to one less than 13 mm ($\frac{1}{2}$ in) thick.

Both tufted and needleloom carpet normally consist of three basic parts; fibres, primary backing, and secondary backing.

The primary backing is the prewoven base fabric into which the pile fibres are tufted by rows of needles on the tufting machine, or needle-punched. The secondary backing is the extra layer often attached to the underside of the primary backing to overcome problems of stretching and buckling and to give added resilience, or body.

When the carpet is completed, the primary backing is generally completely hidden between the surface fibres and the secondary backing underneath.

The primary backing may consist of woven jute or a synthetic fibre, for example polypropylene. The secondary backing may also consist of jute or a synthetic material, similar to those used to form the surface fibres. Alternatively, foam rubber is sometimes used.

Carpets constructed entirely from synthetic materials are impervious to moisture, rot and mildew. Accidental spills will not be absorbed by the primary backing and the carpet can be more thoroughly cleaned than can one made with natural fibres.

Tufted carpets made from synthetic materials can be used in entrance halls and lifts, where wet foot traffic can cause a lot of water to be trampled into the carpet. These carpets are dimensionally stable and remain virtually unaffected by changes in temperature and humidity.

Knitted carpets are manufactured on knitting machines and a backing is attached to give stability and weight. Few knitted carpets are to be found in the U.K., although they are widely used in some areas of the U.S.A.

Pile-bonded carpets consist essentially of fibres, which may be of natural or synthetic origin, compacted or pressed into an adhesive backing. The whole is then heated and dried. Alternatively, synthetic fibres may be bonded by melting so that they fuse together on to a backing. A secondary backing may then be attached to provide stability, if required.

Electrostatically flocked carpets normally comprise a very short nylon staple, often dyed, a primary backing which is usually jute and a secondary backing which may be jute or foam. Flocked carpets are characterised by very dense pile with a uniform surface and are widely used on the floors of motor cars.

The advent of carpet tiles has opened new possibilities for floor coverings. They are to be found in many widely different locations, for example offices, restaurants, shops, banks, churches and other places where a floor covering is not normally laid. Not only can attractive patterns be designed, the tiles can be easily interchanged to even the wear and worn or soiled tiles easily replaced. They also allow comparatively easy access to floor ducts and underfloor services.

Carpet tiles are made from many different fibres. Some consist of pure or crimped wool, others of blends of fibres, for example nylon and animal hair.

Tiles made solely from nylon are usually stuck to the floor. Those made from a blend of synthetic fibre and animal hair are normally loose-laid.

CHARACTERISTICS

Carpet can provide an infinite variation of design, texture and colour, and a more attractive floor than, perhaps, any other type of material. It minimises noise and is extremely quiet to the tread, has good heat insulation properties and is noted for its warmth underfoot.

Resistance to abrasion, or wearing properties, vary widely with the kind of carpet and the type and volume of traffic. A considerable amount of development work is currently in progress to improve even further the durability of almost all types of carpet.

The durability can be considerably extended by the use of an underlay. If the sub-floor on which a carpet is laid is not smooth and level, excessive wear can take place in certain places, particularly in any raised irregular areas. The use of an underlay acts as a shock absorber between the carpet and floor and compensates for any unevenness in the floor itself. An underlay can play a major part in prolonging the life of a carpet, especially if the pile is short, as in a worsted Wilton. Its presence will minimise the effects of an uneven floor and give added resilience and warmth.

An underlay should be laid so that no gaps or overlapping takes place. Either will cause an uneven surface for the carpet, with the possibility of accelerated wear.

Types of underlay include needleloom felt, hair felt, synthetic rubber and various types of synthetic foam materials.

Resistance to chemicals varies widely depending on the type of carpet and fastness of the dyes. If any chemical spillage takes place on a carpet it should be removed immediately.

Damage to some carpets can be cause solely by water, due to spillage, flooding, a leaking water pipe or other means. Unpleasant mildew odours can occur if the carpet backing rots; such odours are sometimes the first indication that water is present.

If a carpet becomes affected by water this should be mopped up with cloths or sponges as quickly as possible. If the carpet is stained, or a water 'tide' mark appears around the affected area, it should be shampooed while still damp. A wet carpet should not be lifted immediately but allowed to remain in place and dried out by all available means. When almost dry it should be lifted and dried thoroughly.

If severe rotting has taken place it may be necessary to cut out and replace the affected area.

A carpet has excellent slip resistance properties and even if someone should fall the possibility of injury is lessened by the softness of the material.

Resistance to indentations caused by the legs of heavy furniture resting on a carpet varies widely, according to its quality and pile. If indentations occur, they can be removed effectively by applying steam to the crushed areas, using a steam iron or a dry iron over a wet cloth. It is important that pressure is not applied on the iron. After steaming, the pile should be brushed up and the operation repeated, if necessary.

Broad furniture glides should be fitted underneath the legs of heavy furniture to minimise the possibility of indentations taking place.

Sunlight can have an adverse effect on some carpets and cause the colours to fade. If possible, carpets should be shaded and turned round, not only to even the wear but also to ensure that if the colours are affected by light a uniform change takes place.

With regard to special characteristics, some carpets can be affected by moths and beetles and loss in the United States alone has been estimated at approximately £30 million annually.

Moths are found in most houses and commercial buildings all year round. They can crawl into wall and floor apertures and remain hidden and unaffected by normal cleaning operations. Moth proofing is essential if damage from this source is to be eliminated.

Static electricity on carpets can present problems. When two different insulating materials come together and are then separated, electrostatic charges accumulate on each. Static is, therefore, formed when foot traffic passes across a carpet, as both shoe soles and carpet fibres are insulators. Friction, also caused by foot traffic, can increase the amount of static.

Static electricity is more prone to build-up on a synthetic carpet than on a natural one. A wool carpet retains a far higher percentage of moisture than an equivalent carpet made from synthetic fibres. Moisture conducts electricity and for this reason any charge formed on a wool carpet normally passes through the carpet to earth. Problems of static electricity are, therefore, rarely, if ever, experienced on wool carpets.

The charge formed on the surface of a synthetic fibre carpet tends to accumulate, either on the carpet or on the person walking on it. When the person concerned touches an earthed object, the electricity is discharged and an electric shock is experienced, which can be extremely unpleasant.

A secondary factor contributing to a build-up of static electricity is a dry atmosphere. If the relative humidity is low, for example in rooms centrally heated and double glazed, static electricity is more liable to occur than under conditions of high humidity. It can also cause carpets to soil quicker than would otherwise happen, as dust can be accumulated as a result of electrostatic attraction. In practice, however, the electrostatic attraction of dust is of little significance, because most dirt and soilage are carried on to the carpet by foot traffic.

Several methods are available for minimising static electricity. If the relative humidity is maintained at about 60 to 75 per cent any static formed will leak away into the atmosphere and thence to earth. Controlling the humidity can only be a temporary solution, however, and anti-static agents may also be necessary to reduce static to an acceptable level.

Some anti-static agents are chemical compounds which attract moisture from the atmosphere, so that the relative humidity in the area of the carpet pile is high. This enables static present to leak away, and their use may be essential on carpets prone to static. They are mostly applied to the carpet by a mechanical spray, although some are available in aerosol form.

It should be recognised, however, that many anti-static agents have limitations. Because of their moist nature, they tend to hold dirt and soilage, with the result that the carpet may lose its clean appearance quicker than if it had not been treated. More frequent cleaning may become necessary. With most anti-static agents, the effects are lost each time the carpet is shampooed, and the agent must be re-applied after each cleaning operation if static electricity is to be controlled.

Some of the most recently developed anti-static agents combine good static proofing qualities with improved resistance to soilage and may remain effective for periods up to about one year.

Anti-static agents are sometimes applied to a carpet during the manufacturing process, but it should be recognised that re-treatment will be necessary in due course and that the initial application will not remain effective indefinitely.

Carpets are also made which incorporate extremely fine metal fibres in the pile to aid conductivity. These are sometimes known as 'non-static' carpets. Others contain metal woven into the backing.

Carpet shampoos help reduce significantly the level of static, but their effectiveness diminishes rapidly and further treatment may be necessary after a comparatively short time.

Perhaps the problems associated with static electricity can best be overcome by the discriminate use of an anti-static agent in conjunction with a carpet shampoo, at a controlled level of relative humidity.

Pile carpets and rugs possess a number of special characteristics which may give rise to concern but are not, in themselves, defects. Perhaps the most important amongst these are known as sprouting, shooting, fluffing, shading, discolouring and shrinkage.

Sprouting is a phenomenon associated with loop pile carpets and is the appearance of a loop of yarn above the surface. It can be caused by a shoe nail, for example, which may pull a loop and thereby cause the adjacent loop to be drawn beneath the surface of the pile. The remedy is to hook up the lowered loop to the surface, thereby pulling back the raised loop. This can be done with a pin or crochet hook. Alternatively, if a loose end of yarn appears above the surface, indicating that the loop has broken, it should be cut off and not pulled, otherwise further loops could be affected.

Shooting is similar to sprouting, and refers to fibres which protrude above the normal surface of a carpet. This fibre should be cut off and not pulled.

Fluffing is a normal characteristic associated with a new carpet. When a carpet is made, some lint or short fibres are left in the pile. Over a period of a few months they slowly rise to the surface and their appearance is called 'fluffing'. The carpet is in no way damaged by their appearance, although they may make it look unsightly. Any excess fibres should be removed by brushing or vacuuming, but vigorous action should be avoided as it could cause an unnecessary loss of material.

Shading is a characteristic of good quality cut pile carpets and refers to an apparent change in colour. It is sometimes very noticeable with plain, light coloured carpets. After a carpet has been laid a short time light and dark patches may appear. This is because the pile lies at different inclinations, so that light is reflected to varying degrees according to whether it strikes the sides or ends of the tufts of pile. The sides reflect light to a greater extent and therefore appear lighter in colour. Areas appear light or dark, depending from where they are seen.

Shading can be minimised by vacuuming to raise the pile and by changing the carpet around, or altering the furniture and fittings so that different traffic lanes are used. Shading can sometimes be lessened by relaying the carpet so that the pile leans away from the light source, whether a window or electric light.

CARPET GROUP OF FLOOR COVERINGS

Discolouring may be due to settlement of dust from the atmosphere that is taking place continuously. Atmospheric dust gives a grey tone to the carpet and can be removed by shampooing to restore colours to their original brilliance. Discolouring may also be due to fading. While the dyes used to give colour to carpet fibres are as fast to light as possible, some fading may take place, particularly if carpets are exposed to direct sunlight for any length of time. Direct sunlight should be avoided by the use of drapes and blinds.

Shrinkage of carpet fibres and backing can happen if too much water is applied, particularly during a shampooing operation. Many fibres used in the manufacture of carpets tend to shrink when wet and, in general, they are not pre-shrunk. A carpet should be fitted to allow for possible shrinkage after cleaning. Although a carpet can be stretched, the extent to which this can be done is limited as it is imperative that the fibres should not be damaged.

Fibres with a tendency to shrink, therefore, should be cleaned with the minimum of water.

MAINTENANCE

When properly maintained, a good quality carpet will give very many years of excellent service and the life of a poorer quality carpet can be considerably extended. If, however, a carpet is neglected, its length of life can be shortened appreciably.

Different types of carpet require different methods of maintenance, and experience has shown that a cleaning procedure ideal for one type may have little, if any, effect on another. It should be recognised that the best method of maintenance for any particular carpet will depend, among other things, on its type, colour and location, as well as on the volume of traffic passing over it and the standard of cleanliness required.

For example, under normal circumstances water should not be allowed to come into contact with a conventional type of carpet unless absolutely necessary. Tiles composed predominantly of hair, however, should be sprayed with water at infrequent intervals to enable the hair to retain its natural moisture. This ensures that the hair remains flexible and extends the useful life of the tiles.

In recent years a great deal of development work has been carried out not only on carpets themselves, but also on cleaning materials and equipment for maintaining them. New and improved cleaning materials and machines are constantly being developed and new methods for cleaning carpets devised.

CARPET GROUP OF FLOOR COVERINGS

It is the aim of this section to outline in general the main principles concerned with carpet cleaning, rather than to detail every facet at length. Many organisations now specialise entirely on cleaning carpets. If a problem arises which requires skilled and detailed knowledge, it is recommended that such an organisation be approached for specialist advice.

When considering carpet cleaning techniques, it is advisable to take into account the various types of soil liable to be found on a carpet. This is because different types of soil are removed by different materials and methods. Once the type of soil is known, it can be removed quicker and more efficiently than might otherwise have been the case.

In general, the main types of soil are dry particulate dirt, water-soluble soil, solvent-soluble soil and stains.

Dry particulate dirt is probably the predominant type of soilage found on most carpets. It consists of grit, sand, dust and many types of coarse, hard particles. Particulate dirt is liable to shorten the life of a carpet quicker than any other. This is because the small particles of dirt frequently have very sharp and abrasive edges, and if the particles are not removed they collect at the base of the pile fibres and cut into them when foot traffic presses down the carpet pile.

Water-soluble soil includes such items as non-greasy food, sweets and starches.

Solvent-soluble soil includes asphalt, tar, oil and grease. Stains may be caused by a very wide range of substances, for example alcohol, coffee and ink.

Dry particulate dirt is generally carried on to the floor by foot traffic. Water- and solvent-soluble soils and stains may be carried on to the floor by foot traffic, but are often the result of accidental spillage.

Where carpet has been laid on floorboards and particularly if there are gaps between the boards, air rising through the gaps may carry dirt into the carpet and deposit it in the pile. The result will be the appearance of dark parallel lines on the carpet, which may be particularly noticeable if the carpet is light coloured or plain.

The problem can be solved by laying brown paper beneath the underlay, if present, so preventing air from rising through the carpet immediately above the gaps in the floorboards.

The mechanism of soil removal will depend on the type of soil, although it should be recognised that in normal circumstances several different types will probably be present.

Dry particulate dirt can be removed by mechanical means, for example carpet sweepers and vacuum cleaners. Water-soluble and solvent-soluble soils can be removed by chemical means, normally using a carpet shampoo. Stains can be removed by a variety of means, including the use of special chemical spot and stain removers.

CARPET GROUP OF FLOOR COVERINGS

To be effective, a carpet maintenance programme must provide for the removal of all types of soil, including dry particulate dirt, water- and solvent-soluble soils and stains.

The frequency with which each cleaning operation should be carried out will vary widely, depending on a number of factors, of which the most important are the location of the carpet, the type and volume of traffic, the colour of the carpet and the standard required. A carpet located in a main entrance hall or heavily used corridor will require more frequent attention than a similar one in an interior area. Also, the frequency of cleaning may need to be increased if the carpet is light coloured and readily shows dirt, or if a particularly high standard is required.

The main processes in carpet maintenance, therefore, include the use of mechanical equipment for the removal of dry particulate dirt, and chemical preparations for removal of water-soluble and solvent-soluble soils and stains.

Use of Mechanical Equipment

Two main methods are available for removing dry particulate dirt from carpets. The first involves the use of a manual, push-pull carpet sweeper and the second an electrically operated vacuum cleaner.

Carpet sweepers are very useful for removing medium and small sized litter and soil. They are widely used in domestic locations and on occasions commercially in light traffic areas.

Dry soil embedded in carpet fibres needs stronger action to dislodge it and in most industrial and commercial locations electric vacuum cleaners are preferred.

Many different types and sizes of vacuum cleaner are available and it is important that the appropriate machine be used for the carpet being cleaned. A small, domestic model, for example, may not be powerful enough to remove coarse grit trodden down to the base of the pile. Machines designed for industrial and commercial use are normally more powerful than domestic types and will remove deeply bedded grit and dirt effectively, as well as surface soilage.

One of the main types of vacuum cleaner relies solely on suction to remove dirt. It consists of a reservoir, or tank, usually on castors, connected by a flexible hose to a carpet tool, which is placed on the floor and pushed forwards and backwards over the carpet to pick up the dirt.

This type is extremely suitable for removing light soilage. It is also very effective on certain types of low pile carpet.

CARPET GROUP OF FLOOR COVERINGS

Another principal type of vacuum cleaner consists of a single, upright, piece of equipment, with a beater brush in the head of the machine in contact with the carpet and a reservoir, or bag, to hold the soil fixed alongside the handle. The action of this type on a conventional pile carpet is to raise the carpet very slightly by suction, so that the pile is in contact with the beater brush. The soil and loose particulate dirt is loosened by vibration and the beater action. As the carpet is raised very slightly off the floor, a strong inward air current is allowed to force the loosened soil and dirt out of the carpet and into the bag.

The beater brush type of vacuum cleaner, however, is frequently found to give unsatisfactory results on needle felt, or non-pile, carpet. On these surfaces, no pile is present and carpet is not raised by suction to enable the beater brush to operate efficiently.

Needle felt or non-pile carpets should be cleaned with a suction type of vacuum cleaner, or a vacuum cleaner adapted so that the beater brush can be lowered on to the surface, where it can operate efficiently.

Some types of carpet tile, particularly those consisting of about 95 per cent hair and 5 per cent synthetic fibres, should be vacuum-cleaned diagonally across the tiles to assist the pile of neighbouring tiles to bind together.

In hospitals, filtering of the exhaust air is equally as important as the ability of the vacuum cleaner to remove soil from the carpets. If the exhaust is not efficiently filtered, organisms picked up from the floor may be distributed over a very wide area.

The frequency of vacuuming will vary from once or twice a week in light traffic areas to, perhaps, twice daily in heavy traffic areas.

Chemical Preparations

Chemical preparations are used to remove water-soluble and solvent-soluble soils which cannot be removed with a vacuum cleaner. Frequency of cleaning will again vary widely, depending on the amount of soil on the carpet and the degree of cleanliness required. In very general terms, cleaning with a chemical preparation may vary from once a month to once a year.

This cleaning can be carried out by sending the carpet to a professional carpet cleaning organisation, or on site. If the carpet is lifted and sent away, problems have occasionally been experienced when re-fitting due to shrinkage. These problems have now been largely overcome by the use of improved cleaning materials and techniques.

Off-site methods of carpet cleaning include deep steam cleaning, a process in which soil is flushed out of the carpet and backing by a controlled jet of hot water and steam sometimes in conjunction with a liquid shampoo. This is followed by drying with a powerful vacuum. Deep steam cleaning requires special equipment and is generally done by trained specialists.

Perhaps the four most important types of chemical preparation for on-site carpet cleaning are; dry powder cleaners, liquid shampoos, dry foam shampoos, and spot removers.

Dry Powder Cleaners

Dry powder cleaners are occasionally used to clean carpets, but not to the same extent as are liquid carpet shampoos.

They consist, essentially of absorbent materials, which may include wood sawdust, saturated with solvents and drying agents. The powder is sprinkled on to the affected area and may be worked into the pile with a stiff brush or machine. After the powder has been allowed time to absorb dirt and the solvents have evaporated, the material is vacuumed from the floor.

These cleaners are quick, simple and fairly effective for certain lighter soils. The main advantage is that the carpet can be in use again in the shortest possible time and much sooner than if liquid shampoo is used. There is also no danger of over-wetting the carpet, so that the possibility of damage to the backing is eliminated.

Perhaps the main disadvantage is that a carpet is not cleaned as thoroughly as with a liquid shampoo. This is because while the solvents present in the powder will have some effect on the solvent-soluble dirt, the water-soluble dirt may remain almost entirely unaffected. Removal of all the dry powder is also extremely difficult and even after vacuuming the carpet repeatedly some is likely to remain at the base of the carpet pile. Carpets cleaned with dry powders occasionally have a tendency to re-soil easier than if cleaned with a dry-residue type of carpet shampoo.

While the use of dry powders is occasionally expedient, they are not intended to replace the more thorough, periodic, cleaning carried out with a carpet shampoo. They can be used, however, to prolong the period between shampooing operations.

Liquid Shampoos

Basically, liquid carpet shampoos belong to two main types, those similar to washing-up detergents and those of the dry-residue type.

CARPET GROUP OF FLOOR COVERINGS

The former are effective in that they clean the carpet satisfactorily. Because of their nature, however, once the water has evaporated a greasy residue remains, forming a thin film over the surface of the carpet fibres. As soon as foot traffic is allowed on the cleaned carpet, dirt from the soles of shoes becomes trapped in the greasy residue, with the result that the carpet tends to re-soil comparatively quickly.

Problems of re-soiling have been largely overcome with the advent in recent years of the dry-residue type of carpet shampoo.

Dry-residue carpet shampoos contain materials different from those used in conventional washing-up type detergents. A small amount of solvent is frequently included to assist with the removal of solvent-soluble dirt.

When diluted with water and applied to a carpet, the dry-residue carpet shampoo loosens the dirt so that it can be easily removed. On drying, the material drys to a microscopic powder. The carpet is then vacuumed, when dirt and the powder are removed and the pile raised.

By this method no residue remains on the carpet and re-soiling is minimised. The intervals between carpet cleaning operations are, therefore, extended.

Before shampooing, a carpet should be tested to ensure that the colours are fast. This should be done by pouring a little neat carpet shampoo on to a white cloth and rubbing a small section of the carpet in one corner. If no colour appears on the cloth it is safe to continue.

The method of cleaning a carpet is first to vacuum it to remove most of the coarser types of dirt. Greaseproof paper should then be wrapped round and under the legs of furniture too heavy to move. Metal furniture should not be allowed to become wet in case it subsequently rusts and stains the carpet.

The next stage when using a dry-residue type of carpet shampoo is to dilute it with water in accordance with the manufacturer's recommendations. It is then applied by machine, applicator, soft brush or sponge, using the minimum amount of liquid necessary to produce a light lather. The foam is then worked evenly over the area and the process repeated until the whole carpet has been cleaned.

It is essential that excess liquid should not be allowed to come into contact with the carpet. Most dirt is retained in the upper part of the carpet fibre and it is generally unnecessary to soak the lower part as well. If water is allowed to penetrate to the carpet backing it could cause it to rot or shrink.

When the whole carpet has been cleaned it should be allowed to dry, then vacuumed to remove the dirt and fine powder residue and to raise the clean pile. Vacuuming is an important part of the cleaning operation as grit, sand and other types of particulate dirt are removed at the same time.

CARPET GROUP OF FLOOR COVERINGS

Although very effective in cleaning a carpet, it should be recognised that generally at least four hours are required for the carpet to dry and if the pile is deep, up to 24 hours may be necessary.

Dry Foam Shampoos

The dry foam type of carpet shampoo has been developed comparatively recently and is very similar to the dry-residue liquid shampoo. The term dry is not strictly correct, as the foam contains some moisture.

Shampoos of the dry foam type are used in conjunction with a machine constructed specially for the purpose. The shampoo is diluted with water and the solution poured into a pressure tank forming part of the machine. The solution is then converted into a foam by a compressor and the foam is worked into the carpet by revolving brushes, underneath the machine. The action of the brushes is to agitate the foam so that it passes through the carpet pile, removing the dirt in the process.

The minimum amount of liquid is allowed to come into contact with the carpet, so that a quicker drying time is achieved, often as little as 30 minutes. The risk of overwetting a carpet is also greatly reduced. Traffic can, therefore, be allowed on the cleaned carpet after a comparatively short time. While the drying time is reduced, however, cleaning efficiency is likely to be less on heavily soiled areas than that achieved with a liquid dry-residue type of shampoo. This is because water, in itself a good cleaning agent, is kept at an absolute minimum to achieve a faster drying time.

The cleaned carpet should be allowed to dry, then vacuumed to remove dry-residue and dirt remaining on the surface of the carpet.

Some of the larger types of machine have a wet pick-up vacuum forming part of the machine, so that drying and soil removal can be quickened. Using machines of this type, the process of dry foam cleaning can be made into a one-pass operation of the machine.

This method of cleaning is very effective for large areas of carpet, where the use of a special carpet cleaning machine is justified. It should be recognised that with regard to effectiveness, however, the dry foam method has some limitations.

Spot Removers

Almost all stains are caused by accidents or other mishaps and it is inevitable that sooner or later a carpet will become marked. When this occurs it is generally unnecessary and undesirable to clean the whole carpet. Spot cleaning to remove the stain from the affected area is normally all that is required.

Stains can be caused by many different agencies, for example tea, ink or paint. Many stains can be completely removed by the correct treatment promptly applied. Wrong treatment, however, may fix the stain, causing an irreversible chemical reaction and thereby resulting in a permanent disfigurement. Similarly, any delay may also allow the stain to react chemically with the carpet, with the same result. Prompt action is, therefore, essential to remove as much spillage as possible and prevent it from penetrating into the carpet pile.

The normal method of removing a stain is to blot up as much as possible with a clean, absorbent cloth or blotting paper, working from the outside to the centre to prevent the stain from spreading. When using a stain or spot remover on a carpet, particularly if it is a solvent-based material or alkaline or acidic in nature, it is strongly recommended that a test be first carried out in some out-of-the-way area to ensure that the colours are fast and that there is no detrimental effect on the carpet. Once it is established that the stain remover is safe, it should be applied with a clean cloth, again working from the outside edges to the centre. The surface of the cloth should be changed frequently and the affected area mopped or well blotted to remove the stain from the carpet. The treatment should be repeated, if necessary, until all traces of the stain have been removed. Finally, the area should be wiped dry with a clean cloth.

Some of the more common stains and their methods of removal are detailed at Appendix I.

Silicone Treatments

These treatments are sometimes used to prevent dirt from entering the carpet fibres, particularly if they are absorbent, as, for example, wool. The main purpose is to facilitate routine cleaning, so that the period between shampooing operations is greatly extended.

Silicone is a very effective water repellent and tends to repel both moisture and dirt. It provides each fibre with an invisible film which repels water, dirt and stains, which can easily be removed.

A silicone treatment generally consists of a solution of silicone, which is sprayed on to the carpet. It is usually dry and has developed its full resistance to water and stains after about 12 hours.

While water and stains can be easily removed, it should be recognised that very hot liquids may penetrate the silicone protective coating to some extent. The treatment is not generally recommended for pure white carpets, as some yellowing may subsequently take place.

While silicone treatments generally give good results on dense pile carpets, those with an open pile construction cannot normally be treated effectively.

7

OTHER FLOORS

This heading is intended to embrace those floors which cannot conveniently be included in the groups considered earlier. Many are extremely important, for example plastic seamless floors, the demand for which is increasing rapidly.

Metal floors are considered first. Iron and steel are dealt with together in one section and they are followed by a separate section on aluminium.

A section on glass floors comes next, followed by plastic seamless flooring, which includes epoxy, polyester and polyurethane screeded, self-levelling and decorative floors.

The chapter concludes with a section on anti-static, or conductive, floors.

IRON AND STEEL

Iron and steel are being used increasingly for industrial floors subjected to heavy wear. It has long been recognised that conventional surfaces such as granolithic are not entirely satisfactory under certain conditions, particularly when a dust-free floor is required.

A number of different types of iron and steel are used for floors. The term 'stainless steel' embraces a number of different alloys, each with slightly different characteristics. Mild steel is also used extensively. Iron and steel are employed in various forms, including tiles, which may be solid or backed with concrete, solid plates, open grid types and fine mesh flooring.

Tiles are available with studded, ribbed or plain surfaces and are rectangular in shape, the most popular size being 300 mm (12 in) square. Steel tiles, or anchor plates as they are sometimes known, and solid faced metal cast iron tiles are used for areas liable to suffer severe impact. The solid iron tiles are used for example in food factories, where a very high standard of cleanliness is required. Steel tiles are frequently used in heavy engineering shops and where severe conditions of wear are found.

They are more suitable for dry than wet or oily conditions, as they tend to corrode when wet and to become slippery if oily. Where failure of the floor takes place it is generally found that the fault lies in the sub-floor rather than in the metal itself.

Solid steel or cast iron plates are used in heavy engineering workshops where heavy objects may be dropped on to the floor, or where there are liable to be spillages of metal at high temperature. Mild steel plates are widely used on the floors of breweries and dairies, where abrasion from casks and churns is very considerable. They are also used, as an alternative to tiles, in bakery dough-mixing rooms, where tanks containing dough mounted on small-wheeled bogies subject the floor to very heavy wear.

Steel or cast iron open grid or fine mesh floors are often suspended for use as raised platforms around chemical plant, walkways in engineering shops and fire escape stairs. When cleaning, care must be taken to ensure dirt and soilage do not fall on to people or plant beneath.

Characteristics

Unpainted iron and steel are generally a drab, metallic colour, and are not normally used for floors where a decorative effect is required.

Both iron and steel floors are extremely hard and generally noisy and cold to the tread. They have outstanding durability and resistance to abrasion and impact. The sub-floor is far more likely to suffer damage than the metal itself. They are impervious to oil, grease and weak acid and alkali, but will be harmed by strong acid.

Patterned surfaces are normally slip resistant, but plain steel or iron plates can become slippery if a film of oil, grease or water is allowed to remain on the surface.

Maintenance

Iron and steel floors are generally relatively easy to maintain. With stainless steel floors, however, it is important that there should be no direct contact between the stainless steel and other, less corrosion-resistant, metals. If the floor is poorly drained and water is allowed to remain in contact with both metals at the same time, an electrolytic action could cause accelerated corrosion of the other metals concerned.

Iron and steel tile and solid plate floors should be swept to remove all loose dirt and soilage and scrubbed with the minimum amount of water and the weakest type of detergent necessary to achieve the desired results. Steam cleaning and high pressure jets may be necessary to remove heavy accumulations of soilage, particularly if the flooring is exposed to soil carried in from outside, or to exceptionally dirty conditions. It should, however, be recognised that wet metal floors, in

OTHER FLOORS

addition to being slippery if smooth, are extremely good conductors of electricity. Electric plugs, leads to electrical floor cleaning machinery and the machines themselves, must therefore, be regularly and frequently inspected to ensure that they are in safe working order.

Any rust found on the floor can generally be satisfactorily removed with an abrasive paste cleaner. If little rust is present a fine paste cleaner should be used, but if there is a considerable amount, a coarser grade of paste cleaner may prove more satisfactory.

Open grids and fine mesh flooring should be cleaned with a stiff brush. Open grid floors are often painted with materials formulated to minimise the incidence of rust. The condition of the paintwork should be inspected at regular intervals to ensure that the metal is being properly protected. Fine mesh flooring may require periodic cleaning with a wire brush to remove dirt and soilage retained in the mesh. Scrubbing with a solution of a neutral detergent in water will also help keep fine mesh flooring clean.

Conventional solvent- and water-based seals and waxes should not be used on iron and steel floors. If some form of surface coating becomes necessary, materials specifically designed for use on metal should be used to ensure obtaining the best possible adhesion.

In general, however, iron and steel floors have a very long life and require very little maintenance.

ALUMINIUM

In recent years this has found an increasing use as a flooring material and aluminium fabricators have developed a variety of open tread or grid flooring systems. They are mainly used as raised platforms, walkways or cat walks, both indoors and outdoors. Aluminium has an extremely high strength to weight ratio; it is practically corrosion-resistant and has extremely good resistance to chemical attack.

One of its most important properties is that it does not generate sparks when struck with another metal. For this reason aluminium walkways are frequently used in the chemical industry, particularly near tanks containing flammable solvents.

Aluminium used for flooring is normally mill finished, that is, without any coating applied. It is, however, protected by a thin, tough film of aluminium oxide which forms on the surface as soon as the metal is exposed to air. If the film becomes scratched, perhaps by foot traffic, it immediately reforms. The film can, however, be attacked by acids. Even weak acids, formed when sulphur dioxide and hydrogen chloride gases in the atmosphere come into contact with water held by soilage on the surface, will attack aluminium and may cause it to become pitted at weak points. Where this occurs there is a very slight corrosion of the metal, but the products of corrosion generally seal the pits so that the amount of corrosion is negligible.

In severe industrial environments where dirt and grime accumulate on the aluminium, a white corrosion product may form due to acid attack. While this can look unsightly, it gives a form of protection and if no action is taken the attack gradually stops.

Maintenance

Aluminium flooring requires very little maintenance, and conventional floor seals should not be used. If some form of protective coating is required a paint system specially formulated for use on aluminium should be used.

The flooring should be brushed regularly to prevent dirt and grime from accumulating in the open treads. Water and a neutral detergent may be used with advantage to remove dirt, if necessary. Greasy or oily deposits may be removed with a solvent-based detergent wax remover, followed by thorough rinsing with clean water.

GLASS

Glass is occasionally used for floors and should not be confused with glazed floor tiles, although the two are often similar in appearance.

Mostly, glass is used in mosaic form and is found in light traffic areas, for example bathrooms, where an attractive appearance is required. Glass mosaics are manufactured by subjecting a mixture of molten glass, coloured pigments and fluxes to an extremely high temperature, after which the mixture is decanted into moulds, allowed to cool and cut to size. They are sometimes produced in hexagonal shapes.

The glass mosaics are generally laid in a screed of sand and cement, as are ceramic tiles. The mosaics are manufactured in a wide variety of bright colours from which very attractive patterns can be made.

The surface is hard and abrasion resistant, but will crack if subjected to sudden impact. It is completely resistant to water and reasonably resistant to most chemicals. It is noisy underfoot and cold.

Maintenance

Newly-laid glass mosaic may be stained with cement on the surface and this should be removed as soon as possible after laying, with a damp cloth or a nylon web pad and a little neutral detergent in water.

OTHER FLOORS

Regular maintenance should consist of brushing, to remove light dirt, followed by mopping with water containing a little neutral detergent. Care should be taken to prevent the grout between the mosaic tiles from becoming damaged, as incorrect maintenance materials and methods can be injurious. Grout can be attacked by acids and harsh alkaline detergents and these materials should, therefore, be avoided. A mild abrasive only should be used, should any be required, as coarse abrasives could damage the grout. Care should be taken when using a mild abrasive to ensure the glass mosaic is not scratched. The abrasive is best applied with a damp cloth or, if necessary, a nylon web pad. Metal fibre floor pads should not be used as fragments left on the floor might rust and stain the grout.

Soap is not recommended as deposits of scum could be left on the floor and cause slip.

It is generally unnecessary to seal or wax glass mosaic floors and the use of these materials is not recommended.

When kept clean by washing, these floors will remain in a bright and attractive condition for very many years.

PLASTIC SEAMLESS FLOORING

The term plastic is used to distinguish between seamless floors laid with synthetic resins as the main ingredients, and those with essentially naturally occurring materials, such as concrete, asphalt and magnesite.

While seamless floors based on naturally occurring materials have been laid for many years, those with synthetic resins have been developed only in recent times. Plastic seamless floors first came into prominence in the late 1950s and have since grown in importance.

Plastic seamless floors are known by many names, including 'poured floors', 'floor toppings', 'resinous floors', 'trowelled plastic floors' and 'floors from cans', the last because the materials from which the floors are made are frequently supplied in cans.

The floors were first developed mainly for industrial use because of their hardness and excellent chemical resistance. Their scope has been widely increased, however, by the addition of decorative floors which look extremely attractive, as well as being functional.

Three main types of synthetic resin are used in the manufacture of plastic seamless floors, namely epoxy, polyester and polyurethane, although other resins are also occasionally used. Seamless floors based on epoxy resins were developed first and those with polyurethane are the most recent.

Plastic seamless floors should not be confused with surface seals, also produced from epoxy and polyurethane resins and widely used for many

OTHER FLOORS

years. Seals produce a surface coating to prevent dirt, stains and foreign matter from entering the floor; to extend the life of the floor and to facilitate subsequent maintenance. They are not considered as flooring surfaces, but rather as semi-permanent materials for application to a floor already in existence.

Plastic seamless floors can be laid on a variety of sub-floors, provided that these are sound in construction and in condition. They are laid on concrete, brick, slate, terrazzo, stone and wood and are generally specified where a high degree of resistance to abrasion and chemicals is required, although the decorative types are also laid for their pleasing appearance.

A number of different methods can be used for laying plastic seamless floors, and maintenance will depend to some extent on the method employed.

Floors intended for industrial wear can be screeded or of the self-levelling type and are generally laid in thicknesses ranging from about 3 mm ($\frac{1}{8}$ in) to over 50 mm (2 in) in some instances. A further decorative type, intended primarily for foot traffic and designed to give an attractive appearance, can be laid in thicknesses ranging from about 2 mm ($\frac{1}{16}$ in) to 6 mm ($\frac{1}{4}$ in). The decorative type of floor is not intended to withstand heavy industrial traffic, or prolonged chemical attack.

Screeded, or trowelled floors as they are sometimes known, consist of resin which may be epoxy, polyester or polyurethane, hardener or accelerator, fillers, pigments and additives, aggregates are also sometimes included. The fillers may consist of silica, quartz, bauxite or a similar material, included to provide body and slip resistance. These are particularly useful in areas where oil spillage is liable to occur and cause slippery conditions, as the angular physical shape of the fillers tends to give them slip retardant properties. Pigments give colour to the floor and the additives are included to provide flexibility and other desirable properties.

Aggregates may be fine or coarse, depending on the type and volume of traffic anticipated.

The raw materials are thoroughly mixed together and poured on to the floor like a sand and cement mix. They are then spread with a steel trowel to produce a level floor.

Self-levelling floors are made with essentially the same raw materials from which the aggregate is sometimes omitted. After mixing, the materials are applied in a thinner, more liquid, state and allowed to flow out evenly without being trowelled. The thickness of self-levelling floors is generally less than that of screeded floors. The result is a smooth, glossy surface intended for light, rather than heavy, industrial traffic.

The decorative type of floor is laid more on the lines of a seal than a screed. A primer is first applied and this is followed by two or more clear flood coats, on which are scattered coloured plastic, marble or granite chips or flakes to produce either a single or multicoloured mottled effect. These are then sanded, if necessary, to flatten the rough edges and finally one, two or more clear glaze coats are applied. The latter protect the coloured chips and provide a smooth, hard wearing surface. Decorative floors of this type are intended for light traffic conditions rather than heavy industrial wear.

Epoxy-screeded and self-levelling floors are used in a very wide variety of situations, for example chemical and food factories, hospitals, laboratories, bakeries, breweries, offices and public buildings. They are particularly suitable for floors liable to be subjected to heavy impact, such as loading docks, aircraft establishments and heavy engineering works.

Decorative epoxy floors are found in shops, offices, hospital entrance halls and corridors, commercial buildings and in domestic locations.

Polyester-screeded and self-levelling floors are used in the same locations as epoxy floors, where excellent chemical resistance and durability are required. They are not, however, found in dry cleaning establishments where chlorinated solvents are used, as polyester is softened by these materials. Citric and acetic acids have a tendency to attack polyester. Such floors, therefore, are not laid in food factories where these acids are liable to be used extensively.

Decorative polyester floors are laid in the same areas as decorative epoxy floors.

Polyurethane-screeded and self-levelling floors are also laid in similar locations as those made with epoxy resins. Decorative polyurethane floors are probably the most widely laid of the plastic seamless floorings and new techniques in formulation and manufacture are constantly being developed. They are growing rapidly in popularity and are best suited in areas subject to light traffic conditions.

Characteristics

Screeded and self-levelling floors are generally laid in a single colour, often red, green, grey, brown or other dark colour. Decorative floors, however, are manufactured in a wide variety of colours, depending on the colour of the plastic or other chips. Single or multi-coloured floors can be laid, with mottled, patterned or random designs to give a very pleasing and attractive appearance.

All the seamless floors are relatively hard and non-porous. By adjustments to the formulation, seamless floors can be given varying desirable properties, for example combining hardness and flexibility.

OTHER FLOORS

The screeded and self-levelling floors are relatively noisy, but the decorative types and particularly polyurethane floors are, perhaps, the quietest of the seamless floors.

Screeded and self-levelling floors have excellent durability, but they have not been in existence sufficiently long to enable an accurate estimate of their life to be obtained. The thicker types of floor will withstand heavy industrial traffic, including metal-wheeled trucks. Many floors are still in an excellent condition after several years heavy wear and clearly have many more years of useful life remaining.

The thinner, decorative types of plastic seamless flooring are also expected to give many years of service.

Where failure of a plastic seamless floor has occurred it has generally been found that the sub-floor, rather than the plastic one, has failed. Plastic seamless floors are normally extremely strong and are often much stronger than concrete.

In general, the chemical resistance properties of plastic seamless floors are excellent and they are frequently specified for this reason.

Epoxy floors are especially effective against attack by alkalis. They will resist oils, fats, greases and most common chemicals and solvents. They also have good resistance to mild acids, but are not as effective under acid conditions as are polyurethane floors.

Polyester floors have good resistance to mild alkalis, oils, fats, greases and most common chemicals and solvents. They have, in general, excellent resistance to acid attack, although some acids, for example citric and acetic may damage them. They are also softened by chlorinated solvents.

Polyurethane floors are noted for their excellent chemical resistance properties, particularly to acid attack. They will resist mild alkalis, oils, fats, greases and the majority of common chemicals and solvents. They are, however, softened to some extent by ketone solvents, for example acetone, but recover their hardness when the solvent has evaporated.

Slip resistance of plastic seamless floors is generally very good and is excellent on those floors including a rough filler or aggregate in the mix. Even smooth floors have good slip resistance properties although slippery conditions can occur if oil or water is allowed to remain on the floor.

The floors have no tendency to dust and are immune to attack by fungi or pests.

Maintenance

Light traffic can be allowed on one of these floors about 24 hours after it has been laid. Hardening will, however, continue for a period ranging from about 5 to 30 days, depending on the type of resin and formulation used. During this period the floor will remain 'tender' and heavy traffic should not be allowed on it.

The manufacturers of some types of plastic seamless flooring, as did the manufacturers of some flexible PVC floor coverings when they were first introduced, claim that maintenance is unnecessary. The majority now recognise, however, that unmaintained floors quickly deteriorate and assume a dirty, scratched and unhygienic appearance. Like any other floor, plastic seamless flooring will wear and become dirty under traffic conditions and so maintenance is essential.

The floor should be swept to remove all loose dirt and soilage and this should be followed by damp mopping with a solution of neutral detergent in water. If necessary, a mild alkaline detergent can be used to remove stubborn soilage.

Rough, industrial floors containing fillers and aggregates will resoil quicker than the smoother, decorative types. More frequent mopping or even scrubbing perhaps with a floor polishing/scrubbing machine, may therefore be necessary.

A water emulsion floor wax, preferably of the buffable type, should be applied, particularly to the decorative floors. The wax will protect the glaze coats, improve the appearance and facilitate routine maintenance by ensuring that dirt is not allowed to scratch and mar the surface. The floor wax should then be maintained by periodic buffing.

Oily sweeping compounds should not be allowed on plastic seamless floors as oily residue may cause slip.

Sealing Plastic Seamless Floors

Epoxy, polyester and polyurethane-screeded and self-levelling floors do not need sealing with either solvent or water-based seals. These floors are extremely hard and non-porous and application of a seal would give little, if any, advantage. It is also very doubtful whether a solvent-based seal would adhere because of the exceptional chemical resistance of these floors.

With regard to the decorative types of flooring, however, the manufacturers frequently recommend that an additional coat of glaze should be applied from time to time. It is inevitable that the glaze coat, protecting the coloured chips, will wear and application of a further coat of glaze will prevent damage to the chips themselves.

Reglazing should be carried out before the surface has worn through to the chips. While the frequency of reglazing will depend on the type and volume of traffic, as a very rough guide once a year should prove sufficient.

OTHER FLOORS

Waxing Plastic Seamless Floors

It has already been mentioned that the surface of plastic seamless floors, particularly of the decorative types, is liable to wear under traffic conditions. When this occurs the floors sometimes look dull and lifeless. Although the manufacturers of these floors sometimes claim that waxing is unnecessary, maintenance with a floor wax will protect the surface, improve the appearance by reducing carbon black heel marks and facilitate routine maintenance.

A solvent-based wax should not be used because of the possibility of slip on these hard floors and a water emulsion floor wax is recommended. For most plastic seamless floors a buffable type is preferred, which can be used in conjunction with spray or foam cleaning maintenance procedures. If the surface is particularly smooth, a dry-bright type will probably give a greater degree of resistance to slip.

Two coats should be applied initially, followed by additional coats as required. Periodically all old wax should be stripped from the floor with a solution of an alkaline detergent in water. The floor should then be well rinsed, adding a little neutralising solution or vinegar to the rinse water and allowed to dry. The water emulsion floor wax should then be re-applied.

Use of Detergents

Plastic seamless floors are not absorbent and will resist repeated cleaning with a detergent and water. Removal of soil is, therefore, relatively easy.

Neutral and mild alkaline detergents are satisfactory for routine cleaning; harsh scouring powders and pastes should be avoided, particularly on the smoother, decorative finishes, as they may scratch and dull the surface. For the same reason, coarse grade metal fibre or nylon web pads should not be used on these surfaces, although they can be used, if required, on the coarser, industrial types of screeded and self-levelling floor. On decorative floors routine cleaning should be done with mopping equipment or a bristle brush, to avoid scratching the surface.

A solvent-based detergent wax remover should be used to remove any spillage of oil or grease as soon as it occurs, so that the floor is not allowed to become slippery.

Soap is not recommended as a routine maintenance material as soapy residue may cause slip.

The use of boiling water or live steam for the washing of floors is not recommended, particularly for epoxy floors. These should be cleaned with a solution of detergent in warm water, followed by a rinse with cold water.

ANTI-STATIC (CONDUCTIVE) FLOORS

While static electricity cannot be seen, its effects can both be seen and felt. Static electricity allowed to build up on a floor can be earthed by someone who touches a metal object such as a filing cabinet or metal reading lamp. The person concerned may experience a mild electric shock, which could be uncomfortable. If the atmosphere is charged with a flammable material, such as ether in a hospital operating theatre, or solvents in a chemical factory, a spark caused by the discharge of electricity could result in not just a mild electric shock, but an explosion or fire. Anti-static floors are therefore used in these areas and also in hospital maternity delivery rooms, x-ray rooms, anaesthetising rooms and in some computer rooms, flour mills and light engineering factories.

Build-up of static electricity can, therefore, become a dangerous hazard. Before considering the maintenance of anti-static, or 'conductive' floors as they are sometimes known, it is important to appreciate how a build-up of static electricity can occur.

Most static electricity is produced by friction. Foot traffic, the movement of wheeled equipment in an operating theatre, removal of clothing or even movement within clothes, particularly nylon, can produce static electricity. As all the moving items are connected in some way to the floor the static electricity tends to build up on the floor surface.

Under normal circumstances electricity discharges slowly to earth and no problems arise. In a dry atmosphere, however, particularly if the humidity in the atmosphere is allowed to fall below about 65 per cent and the floor is insulated with a floor covering, static electricity can start to build up.

Perhaps the best solution to static electricity problems is provision of anti-static or conductive floors. These conduct a specified and controlled amount of electricity to earth, so that the hazards that can be created by a build-up of static electricity are eliminated.

The electrical conductivity of an anti-static floor is normally specified at the time of laying. The electrical resistance is checked, if possible, before the floor is laid, particularly if a sheet or tile type of anti-static floor covering is to be used. The electrical resistance is also tested after laying to ensure that the results lie within the specified range. It is recommended that the floor be tested every three months, or more frequently if necessary, to check that no deterioration has taken place.

OTHER FLOORS

Measurement of electrical resistance should be carried out by a person qualified to undertake this type of work.

It is suggested that a record should be kept of the history of each anti-static floor. Entries should include any maintenance that has taken place, together with all results of electrical resistance tests. By this method any change in anti-static properties can be seen and any long-term trends noted, so that any remedial action needed can be taken in good time.

Several types of anti-static floor have been developed and all are sensitive to slight changes in conditions. They are expensive, not only because installation costs are high, but also because they have a limited life.

Perhaps the most important types of anti-static floor include terrazzo and magnesite, which are laid *in situ* and linoleum, rubber and flexible PVC, which are factory made. Linoleum and rubber are generally manufactured in sheet form. Flexible PVC is available in both sheet and tile form.

Anti-static terrazzo is frequently black in colour due to the inclusion of carbon black in the cement. A galvanised wire mesh is normally laid over the screed to aid conductivity.

Magnesite is given conductive properties by the addition of copper salts. Because magnesite is sensitive to water vapour, the humidity should be maintained at approximately 55 per cent.

One of the difficulties that may be encountered with both anti-static terrazzo and magnesite floors is that because they are laid *in situ,* the uniformity of conductivity will vary and depend, to some extent, on the skill and experience of those responsible for laying the floors.

Anti-static linoleum, rubber and flexible PVC floor coverings have the advantage that they can be fully tested before they are laid. Linoleum may contain carbon to assist electrical conductivity and can be fitted with wire mesh, so that the sheets can be electrically connected by mechanical means when the floor is laid. Anti-static properties in rubber floors are generally achieved by including carbon black in the rubber mix during the manufacturing process.

Sheet flexible PVC may contain carbon, and some types are fitted with thin strips of metal which can be joined together to achieve anti-static properties. PVC tiles are also produced, sometimes with PVC chips, some of which contain conductive carbon. By this method lighter coloured floors can be laid with consistent electrical qualities.

Both PVC sheet and tiles are normally welded at the seams to provide a uniform, continuous surface, without joins or cracks.

In recent years electrically conducting adhesives have been developed which are proving very successful, both in sticking down resilient floors and in providing electrical conductivity.

OTHER FLOORS

While anti-static resilient floors are, in general, not as durable as the *in situ* types, they are much easier to replace if some fault should occur, so that any disruption is minimised.

Characteristics

Anti-static floors, in general, retain the characteristics of their ordinary counterparts, but some special characteristics are apparent.

Colours are often darker, usually because of the presence of carbon black. Durability is frequently reduced, not because the resistance to abrasion is less, but because the electrical conductivity properties may deteriorate. If this falls outside the specification tolerance limits, the floor must either be repaired or replaced, regardless of the amount of wear it has taken.

The presence of oil, grease, wax and similar materials on an anti-static floor may damage the surface itself, or have a detrimental effect on the anti-static properties of the floor. They must, therefore, be avoided.

Maintenance

Correct maintenance of an anti-static floor is absolutely essential. No other type of floor requires greater care. Correct maintenance will not only keep the floor in a clean and hygienic condition, but will also prolong its life and ensure that the conductivity remains within specified limits, a necessary factor if the floor is to be safe and effective.

Unlike conventional floors, it is imperative that protective coatings should not be applied, as they might interfere with the anti-static properties of the floor. Conventional floor seals and waxes, therefore, should never be used, as even very thin films produced by these materials could seriously upset the electrical conductivity of anti-static floors. Similarly, soap should never be used for cleaning purposes, as a thin film might remain and impair the anti-static properties. Even frequent washing with hard water could leave a film, just as lime in water causes deposits to form in water pipes, kettles and similar items.

New terrazzo and magnesite anti-static floors must be brushed thoroughly to remove all light soilage and dirt. The floor should then be cleaned with a liquid neutral detergent solution in water. If necessary, a mild abrasive paste or neutral abrasive powder, containing fine abrasives which will not cause scratching, may be used. The floor should then be thoroughly rinsed and allowed to dry.

Daily maintenance should consist of damp mopping with a neutral detergent solution. Weekly, or more often if necessary, the floor should be cleaned with a fine abrasive powder to remove any stubborn soilage. It is important when dealing with magnesite floors to use only the minimum amount of water necessary, as excess water can have a detrimental effect.

Bactericidal detergents are frequently used on hospital anti-static floors to ensure that the highest possible standards of cleanliness and hygiene are vigorously maintained.

New linoleum, rubber and flexible PVC anti-static floors should be allowed to settle for about seven days after laying to allow the adhesive to harden thoroughly. They should then be maintained with a mild, neutral, liquid or powder detergent in water. It is important that only the minimum amount of water be used, as excess may seep between any joints present and loosen the adhesive. If, however, the seams between sheets or tiles have been welded so that water cannot seep through joints, then as much water can be used as is necessary to obtain a clean and hygienic floor. Bactericidal detergents may be preferred in hospitals, as mentioned above.

Problems sometimes arise with the cleaning of any metal strips in an anti-static floor. It is often very difficult to clean them all to the same extent, with the result that they can present a patchy appearance. Short of cleaning each metal strip individually, a laborious and time-consuming task, there is little that can be done. Provided the electrical conductivity is satisfactory, however, the very slight deterioration in appearance that this represents is generally considered to be acceptable.

When dealing with anti-static floors, the key to success lies in frequent cleaning with the minimum of water and materials. Similarly, frequent cleaning with water will also completely remove or minimise any build-up of static electricity that might occur on other types of floor, such as flexible PVC, where a build-up is liable to take place. Washing with water removes the charge, although it can build up repeatedly if further action is not taken to wash the floor regularly and often, or to increase the humidity of the atmosphere to a minimum of about 65 per cent.

Care should be taken with the maintenance of floors adjacent to anti-static floors. Solvent waxes, particularly, can be transferred on the soles of shoes from one area to another. Wax carried on to an anti-static floor might greatly affect the electrical conductivity of the floor and any deposited thus should be removed immediately. Adjacent floors must, therefore, be taken into account when considering the maintenance of anti-static floors.

APPENDIX I

REMOVAL OF STAINS FROM CARPETS

Reference has already been made of the need for prompt action when dealing with stains on carpets. As much spillage as possible should be removed immediately, by mopping up any liquid with a cloth or blotting paper, or scraping up any solid matter.

Even if most of the stain is removed, a stain or spot remover will probably be needed to remove all traces effectively.

There is no single spot remover that will cope with all stains and the material that should be used will depend on the composition of the stain itself.

Some of the more common stains and their methods of removal are detailed in the following table. If, after treatment, the stain persists, it is recommended that professional advice be obtained.

APPENDIX I

Table A.1. Removal of Stains from Carpets

Stain	Method of Removal
Acid	Apply a solution of bicarbonate of soda, or ammonia, allowing one tablespoonful of the latter per pint of water. Let the solution remain in contact for the least time possible, as the colours may be affected. Rinse with cold water and dry with a clean cloth.
Alcoholic liquids	Apply a solution of a neutral detergent or carpet shampoo and mop or blot with a clean dry cloth. Then sponge with a solution of one part white vinegar to three parts warm water and mop or blot. Finally, rinse with clean, warm water and dry with a clean cloth.
Beer	See *Alcoholic Liquids*.
Beverages (Soft drinks)	See *Alcoholic Liquids*.

APPENDIX I
Removal of Stains from Carpets (contd)

Stain	Method of Removal
Blood	If fresh, blot with a clean cloth and apply cold water. If dried, sponge with a solution of carpet shampoo working from the outside to the centre. Next sponge thoroughly with clean water and dry with a clean cloth. If stain persists, apply a few drops of hydrogen peroxide and after 2 to 3 minutes sponge with cold water and dry with a clean cloth.
Butter	Apply carbon tetrachloride or trichlorethylene with a clean cloth, rubbing from the outside edge towards the centre. Allow to dry for a few minutes, then wipe off with a clean cloth.
Chewing gum	Apply a ring of carbon tetrachloride around the deposit and remove when loosened. Take care that liquid is not allowed to come into contact with any adhesive which may be present in the carpet backing. Alternatively, apply dry ice and scrape off the gum when it becomes brittle.
Coffee	if black, sponge with clean warm water. If white, apply a solution of neutral detergent or carpet shampoo in water and mop or blot with a clean dry cloth. Rinse well with clean warm water. If the stain persists, apply carbon tetrachloride or trichlorethylene with a cloth, rubbing from the outside to the centre. Change the surface of the cloth frequently and mop or blot well between each application. Alternatively, apply a solution comprising equal parts of glycerine and water. Scrub with a stiff brush, and rinse with clean water and dry with a clean cloth.
Crayon.	Sponge with carbon tetrachloride or trichlorethylene and blot dry with a clean cloth. Remove any trace remaining with a solution of carpet shampoo in water. Rinse well and blot dry.
Dye	Remove using glycerine and a clean cloth. If stain persists apply a petroleum jelly with a sponge and finally rinse with clean water. Blot dry with a clean cloth.
Egg	Apply a solution of neutral detergent in warm water and mop or blot with a clean dry cloth. Repeat until all stain is removed and rinse with clean water.
Fat	Apply carbon tetrachloride or trichlorethylene with a clean cloth, rubbing from the outside towards the centre. Allow to dry for a few minutes, then wipe off with a clean, white cloth.

APPENDIX I

Removal of Stains from Carpets (contd)

Stain	Method of Removal
Food	Protein stains should be removed as described under *Blood*. If greasy, treat as *Butter* and if non-greasy apply a solution of carpet shampoo and mop or blot with a clean dry cloth. Finally, rinse with clean water and blot dry.
Fruit juice	Sponge with a solution of carpet shampoo in water. Rinse well with clean water and blot dry with a clean cloth. If the stain persists, apply methylated spirits and mop or blot with a clean cloth.
Furniture polish	Apply carbon tetrachloride or trichlorethylene with a clean cloth, rubbing from the outside towards the centre. Blot dry with a clean, dry cloth. If the stain persists, sponge with a solution of carpet shampoo in water. Rinse well and blot dry with a clean cloth.
Glue, animal	Apply a solution of carpet shampoo in water and mop or blot with a clean cloth. Rinse with clean water and blot dry with a clean cloth.
Glue, plastic	Apply nail polish remover and blot with a clean cloth. If necessary, sponge with a solution of carpet shampoo in water. Rinse well and blot dry with a clean cloth.
Gravy	Scrape up as much as possible with a spoon. Sponge with a solution of carpet shampoo in water and mop or blot with a clean cloth. Rinse well and blot dry. If a grease stain remains, apply carbon tetrachloride or trichlorethylene with a clean cloth and blot dry.
Grass	Apply nail polish remover or methylated spirits with a clean cloth. Blot dry with a clean dry cloth. If the stain persists, sponge with a solution of carpet shampoo in water and mop or blot with a clean cloth. Rinse well and blot dry.
Grease	Apply carbon tetrachloride or trichlorethylene with a clean cloth, working from the outside towards the centre. Blot dry with a clean cloth. Then apply a solution of carpet shampoo in water and mop or blot with a clean cloth. Rinse well and blot dry. If the stain is bad further treatment may be necessary later.
Hair cream	Apply a solution of carpet shampoo in water. Rinse well and blot dry with a clean cloth.
Hand cream	Treat as *Hair Cream*.

APPENDIX I
Removal of Stains from Carpets (contd)

Stain	Method of Removal
Ice cream	Sponge with a solution of carpet shampoo in water and mop or blot with a clean cloth. If stain persists, apply carbon tetrachloride or trichlorethylene with a clean cloth. Rinse well and blot dry.
Ink, ballpoint	Apply methylated spirits with a clean cloth. Blot dry. Sponge with a solution of carpet shampoo in water, rinse and blot dry with a clean cloth.
Ink, copying	Apply methylated spirits with a clean cloth. Blot quickly to prevent stain from spreading. Repeat, then apply a solution of a neutral detergent in water, rinse and mop or blot with a clean cloth.
Ink, writing	Apply a solution of neutral detergent in water, blotting quickly to prevent the stain from spreading. Rinse well and mop or blot dry with a clean cloth.
Jam	Sponge with a solution of carpet shampoo in water, rinse well and mop or blot dry with a clean cloth.
Marmalade	Treat as *Jam*.
Milk	Sponge with a solution of carpet shampoo in water and mop or blot dry with a clean cloth. If the stain persists, apply carbon tetrachloride or trichlorethylene with a clean cloth. Rinse well and blot dry.
Mud	Allow to dry, then brush off loose dirt, or vacuum. Apply a solution of carpet shampoo in water and mop or blot with a clean cloth. Rinse well and again mop or blot dry.
Nail varnish	Apply nail polish remover, then sponge with a solution of neutral detergent or carpet shampoo in water. Rinse well and mop or blot dry with a clean cloth.
Oil	Treat as *Grease*.
Paint, oil based	Apply white spirit (turps substitute), then sponge with a soap solution and rinse well with warm water. Blot dry with a clean cloth. If the stain persists, seek the paint manufacturer's advice.
Paint, water based	Apply a solution of neutral detergent or carpet shampoo in water, rinse well and mop or blot with a clean cloth.

APPENDIX I
Removal of Stains from Carpets (contd)

Stain	Method of Removal
Rust	Apply a solution of neutral detergent or carpet shampoo and mop or blot off with a clean cloth. Rinse well. If stain persists, prepare a solution of oxalic acid (take care, poisonous) and apply with a clean cloth. Hands should be protected with rubber gloves. Rinse repeatedly, adding a little ammonia to the first rinse water. Finish with clean water and mop or blot dry.
Shoe polish	Apply white spirit (turps substitute) with a clean cloth, turning the cloth frequently. If the stain persists, apply carbon tetrachloride or trichlorethylene, then blot dry. Sponge with a solution of carpet shampoo in water. Rinse well and blot dry with a clean cloth.
Spirits	See *Alcoholic liquids*.
Sugar	Sponge with warm water. If a sugary stain persists, apply soap solution, rinse and mop or blot with a clean cloth.
Tar	Apply carbon tetrachloride or trichlorethylene and dry with a clean cloth. Then apply a solution of carpet shampoo in water, rinse well and mop or blot with a clean cloth.
Tea	Sponge with a solution of carpet shampoo in water, rinse well and mop or blot dry with a clean cloth. If the stain persists, apply carbon tetrachloride or trichlorethylene, changing the surface of the cloth frequently. Rinse well and blot dry.
Urine	Mop and blot as much as possible immediately. Sponge thoroughly with warm water, ensuring that liquid does not penetrate the back of the carpet. Apply a solution of carpet shampoo to which has been added white vinegar, in the proportion of one part white vinegar to four parts shampoo solution. Sponge and rinse well with cold, clean water. Mop or blot dry with a clean cloth.
Vomit	Sponge with a solution of carpet shampoo or neutral detergent in water. Rinse well with clean hot water and mop or blot dry with a clean cloth.
Wax	Remove as much as possible by scraping. Then cover with blotting paper and apply a warm iron. The melted wax is absorbed by the blotting paper. Apply white spirit (turps substitute), carbon tetrachloride or trichlorethylene with a clean cloth, changing the surface of the cloth frequently.
Wine	See *Alcoholic liquids*.

APPENDIX II

COVERAGE OF FLOOR SEALS AND WAXES

It should be recognised that the coverage obtained with a product on any particular floor depends on a number of factors, the most important being perhaps the porosity of the floor, the consistency of the product being applied, and the skill and experience of the operator.

The coverage of a floor seal or floor wax can be reduced greatly if the floor is porous and absorbent. If, for instance, a solvent-based seal is applied to unsealed wood or cork, the second and any subsequent coats will normally give a greater coverage than the first. Similarly, if a water emulsion floor wax is applied to a porous linoleum or thermoplastic tile floor, the coverage of the first and even second and third coats may be very greatly reduced. In these circumstances, a water-based seal applied to the bare floor will seal the pores, reduce the number of coats required to provide a satisfactory finish and greatly extend the coverage of each subsequent coat of floor wax.

Floor seals are mostly supplied ready for use, without the need for adding further solvent. Two-pot seals are normally formulated so that once the base and accelerator are mixed the material is ready for application, without addition of solvent.

The viscosity, or resistance to flow of a floor seal, is specified during the formulation stage and carefully controlled during manufacture. The viscosity will determine the ease with which the seal can be spread when applied to the floor. If it spreads too easily, a very high coverage will be obtained, but the protective film of seal remaining on the surface will be very thin, probably resulting in poor durability. If the seal is difficult to spread, the result will be a poor rate of coverage and too thick a film of seal, perhaps resulting in slow drying or other film defects. For optimum results, therefore, the seal should be applied so that the specified coverage, resulting in correct film thickness, will be obtained.

In very cold weather the viscosity of some solvent-based seals, for example oleo-resinous seals, can increase to a marked degree, so that application is difficult and the specified coverage is not obtained. Under such conditions it is recommended that before seal is applied it should be warmed gently, perhaps by placing the container adjacent to some warm central heating pipes, or in a warm room, so that on application it flows easily.

APPENDIX II

Unskilled or inexperienced operators are more likely to obtain excessive or too little coverage, either of which could be detrimental to the seal or floor wax being applied.

Table A.2 is intended only as a guide and is compiled from practical experience gained over a period of many years. It could prove useful when estimating quantities required for any particular situation.

APPENDIX II

Table A.2. Coverage of Floor Seals and Waxes, Part 1 – Floor Seals
Approximate Coverage in m²/litre (ft²/gal)

Type of Floor	Clear Seals					
	Oleo-resinous	One-pot Plastic		Two-pot Plastic		
		Urea-formaldehyde	Polyurethane	Urea-formaldehyde	Polyurethane	
Hardwood	12 (600)	16 (800)	14 (700)	14 (700)	14 (700)	
Softwood	10 (500)	14 (700)	12 (600)	12 (600)	12 (600)	
Wood Composition	10 (500)	14 (700)	12 (600)	12 (600)	12 (600)	
Cork	10 (500)	14 (700)	12 (600)	12 (600)	12 (600)	
Magnesite	10 (500)	14 (700)	12 (600)	12 (600)	12 (600)	
Concrete					12 (600)	
Granolithic					12 (600)	
Cork Carpet	10 (500)	14 (700)	12 (600)	12 (600)	12 (600)	

APPENDIX II

Table A.2. Coverage of Floor Seals and Waxes, Part 1 — Floor Seals (contd)
Approximate Coverage, m^2/litre (ft^2/gal)

Type of Floor	Pigmented Seals		Water-based Seal	Silicate Dressing
	One-pot Synthetic Rubber	Two-pot Polyurethane		
Magnesite	6 (300)	10 (500)		
Concrete	8 (400)	12 (600)		7 (350)
Granolithic	8 (400)	12 (600)		7 (350)
Terrazzo			80 (4 000)	
Marble			80 (4 000)	
Natural Stone			80 (4 000)	
Clay Tiles			80 (4 000)	
Brick			72 (3 500)	
Cement latex			72 (3 500)	
Asphalt	10 (500)	12 (600)	80 (4 000)	
Linoleum			80 (4 000)	
Thermoplastic tiles			80 (4 000)	
PVC (vinyl) asbestos			80 (4 000)	
Flexible PVC			80 (4 000)	
Rubber			80 (4 000)	
Plastic seamless			80 (4 000)	

Table A.2. Coverage of Floor Seals and Waxes, Part 2 – Floor Waxes
Approximate Coverage, m^2 (ft^2)

Type of Floor	Solvent Waxes		Water Waxes	
	Liquid Wax 1 litre (1 gal)	Paste Wax 1 kg (1 lb)	Water Emulsion Floor Wax 1 litre (1 gal)	Foam Aerosol 482 g (1 lb 1 oz)
Wood group (unsealed)	60 (3 000)	61 (300)		
Wood group (sealed)	60 (3 000)	72 (350)	80 (4 000)	372 (4 000)
Concrete (sealed)			80 (4 000)	372 (4 000)
Granolithic (sealed)			80 (4 000)	372 (4 000)
Terrazzo			100 (5 000)	465 (5 000)
Marble			100 (5 000)	465 (5 000)
Natural Stone			100 (5 000)	465 (5 000)
Brick			80 (4 000)	372 (4 000)
Cement latex			80 (4 000)	372 (4 000)
Asphalt group			100 (5 000)	465 (5 000)
Linoleum	60 (3 000)	72 (350)	100 (5 000)	465 (5 000)
Cork Carpet	60 (3 000)	61 (300)	80 (4 000)	372 (4 000)
Thermoplastic tiles			100 (5 000)	465 (5 000)
PVC (vinyl) asbestos			100 (5 000)	465 (5 000)
Flexible PVC			100 (5 000)	465 (5 000)
Rubber			100 (5 000)	465 (5 000)
Plastic seamless			100 (5 000)	465 (5 000)

APPENDIX III

FLOOR MAINTENANCE CHART

Earlier chapters have dealt in some detail with various types of industrial floor and methods of maintenance. It has been stressed that discrimination in selection of floor maintenance materials is essential if the best possible results are to be obtained. Correct materials can give excellent results, whereas others can damage a floor permanently and must not be used.

Table A.3 is intended to summarise the detergents, seals and waxes which should be used and those which should be avoided.

With regard to detergents, neutral and weak alkaline materials are normally safe for almost all floors, although for some, such as wood, water should be kept to a minimum. Strong alkaline materials are also safe for most floors if used sparingly for specific cleaning or wax stripping, but are not generally recommended for normal daily cleaning.

A deterioration in appearance can occur if floor wax is allowed to build up over a period of time. It will probably be a very gradual process and not always apparent.

It is recommended that periodically a small area of floor should be spot cleaned, so that by comparison the degree of any deterioration in cleanliness can be assessed.

On the wood group of floors spot cleaning is best carried out with a clean cloth and a solvent-based detergent wax remover. On other types of floor a mild-abrasive paste cleaner or solution of an alkaline detergent should be used.

When considering floor seals, the importance of correct preparation cannot be over-emphasised, as the durability of any seal depends to a large extent on its adhesion to the floor.

Floor waxes are rather simpler, as the choice lies essentially between water- and solvent-based materials. It should be recognised that when a floor wax is applied over a seal it does not come into contact with the floor itself. For this reason water emulsion floor waxes are sometimes used to maintain wood and similar floors. Thought should, however, be given to the situation that may arise when the seal eventually starts to wear and no longer forms a protective layer between wax and floor.

APPENDIX III

Table A.3. Floor Maintenance Chart. Part 1 – Detergents

Type of Floor	Detergents to Use	Detergents to Avoid
Wood Wood composition Cork	Solvent-based detergents Neutral detergents	Alkaline detergents Abrasive powders Detergent crystals
Magnesite	Solvent-based detergents Neutral detergents	Strong alkaline detergents Abrasive powders Detergent crystals
Concrete Granolithic	Neutral detergents Alkaline detergents Detergent crystals Solvent-based detergents	None
Terrazzo Marble	Neutral detergents Mild alkaline detergents Mild abrasive powders	Strong alkaline detergents Detergent crystals Solvent-based detergents Oily materials

Natural stone:
  Granite
  Limestone
  Sandstone
  Quartzite
  Slate
	Neutral detergents
	Alkaline detergents
	Abrasive powders
	Detergent crystals
	Solvent-based detergents
		Oily materials

Clay (quarry) tiles	Neutral detergents Alkaline detergents Mild abrasive powders	Solvent-based detergents Oily materials
Brick	Neutral detergents Alkaline detergents Solvent-based detergents	Oily materials
Cement latex	Neutral detergents Mild alkaline detergents	Solvent-based detergents Oily materials
Mastic asphalt Pitch mastic	Neutral detergents Alkaline detergents Detergent crystals	Solvent-based detergents Oily materials
Linoleum	Neutral detergents Mild alkaline detergents Solvent-based detergents	Strong alkaline detergents Abrasive powders Detergent crystals
Cork carpet	Neutral detergents Solvent-based detergents	Strong alkaline detergents Detergent crystals
Thermoplastic tiles PVC (vinyl) asbestos Flexible PVC Rubber	Neutral detergents Mild alkaline detergents	Strong alkaline detergents Oily materials Detergent crystals Abrasive powders (on flexible PVC)

APPENDIX III

Table A.3. Floor Maintenance Chart. Part 1 – Detergents (contd)

Type of Floor	Detergents to Use	Detergents to Avoid
Iron and steel	Neutral detergents Solvent-based detergents	None
Aluminium	Neutral detergents Solvent-based detergents	Strong alkaline detergents
Glass	Neutral detergents Mild alkaline detergents Mild abrasive powders	Strong alkaline detergents Coarse abrasive powders Oily materials

Plastic seamless:
  Screeded
 	
  Self-levelling
  Decorative
	Neutral detergents
	
	Alkaline detergents
	Solvent-based detergents
		Abrasive powders (on decorative floors)
		Oily materials

Anti-static:
  Terrazzo
  Magnesite
  Linoleum
  Rubber
  Flexible PVC
	Neutral detergents
	Mild alkaline detergents
	Fine abrasive powders
		Strong alkaline detergents
		Detergent crystals
		Oily materials

APPENDIX III

Table A.3. Floor Maintenance Chart. Part 2 – Floor Seals

Type of Floor	Floor Seals to Use	Floor Seals to Avoid
Wood Wood composition Cork	Solvent-based clear seals	Water-based seals
Magnesite	Solvent-based clear seals Water-based seals (coloured if required)	Silicate dressing
Concrete Granolithic	One- and two-pot polyurethane clear and pigmented seals Synthetic rubber pigmented seal Silicate dressing Water-based seals	Conventional seals liable to be affected by alkali
Terrazzo Marble	Water-based seals	Solvent-based seals Silicate dressing

Natural stone:
  Granite
  Limestone
  Sandstone
  Quartzite
  Slate
	Generally none; but water-based seals can be used on some floors if necessary	Solvent-based seals

Clay (quarry) tiles	Generally none, but water-based seals can be used if necessary	Solvent-based seals
Brick Cement latex	Generally none, but water-based seals can be used if necessary	Solvent-based seals
Mastic asphalt Pitch mastic	Water-based seals (coloured if required) Two-pot polyurethane pigmented seals Synthetic rubber pigmented seal	Solvent-based clear and pigmented seals (except two-pot polyurethane and synthetic rubber seals) Silicate dressing
Linoleum	Water-based seals	Solvent-based seals Silicate dressing
Cork carpet	Generally none, but solvent-based clear seals can be used if necessary	Water-based seals Silicate dressing
Thermoplastic tiles PVC (vinyl) asbestos Flexible PVC Rubber	Water-based seals	Solvent-based seals Silicate dressing

APPENDIX III

Table A.3. Floor Maintenance Chart. Part 2 – Floor Seals (contd)

Type of Floor	Floor Seals to Use	Floor Seals to Avoid
Iron and steel Aluminium	None	Conventional solvent and water-based seals
Glass	None	Solvent-based and water-based seals

Plastic seamless:
  Screeded
  Self-levelling
  Decorative
	None	Conventional solvent-based seals
		Silicate dressing

Anti-static:
  Terrazzo
  Magnesite
  Linoleum
  Rubber
  Flexible PVC
	None	Solvent-based and water-based seals
		Silicate dressing

APPENDIX III

Table A.3. Floor Maintenance Chart. Part 3 – Floor Waxes

Type of Floor	Floor Waxes to Use	Floor Waxes to Avoid
Wood Wood composition Cork	Solvent-based waxes (N.B. If floor is well sealed a water emulsion floor wax can be used)	If unsealed, avoid water emulsion floor waxes
Magnesite	Solvent-based waxes (N.B. If a floor is sealed a water emulsion floor wax can be used)	If unsealed, avoid water emulsion floor waxes
Concrete Granolithic	Water emulsion floor waxes Solvent-based waxes	None
Terrazzo Marble	Water emulsion floor waxes	Solvent-based waxes

Natural stone:
  Granite
  Limestone
  Sandstone
  Quartzite
  Slate
	Generally none, but water emulsion floor waxes can be used if necessary	Solvent-based waxes

Clay (quarry) tiles	Generally none, but water emulsion floor waxes can be used if necessary	Solvent-based waxes
Brick Cement latex	Generally none, but water emulsion floor waxes can be used if necessary	Solvent-based waxes
Mastic asphalt Pitch mastic	Water emulsion floor waxes (coloured if required)	Solvent-based waxes
Linoleum	Solvent-based waxes Water emulsion floor waxes	None
Cork carpet	Solvent-based waxes	If unsealed, avoid water emulsion floor waxes
Thermoplastic tiles PVC (vinyl) asbestos Flexible PVC Rubber	Water emulsion floor waxes	Solvent-based waxes
Iron and steel Aluminium	None	Solvent-based waxes and water emulsion floor waxes

APPENDIX III

Table A.3. Floor Maintenance Chart. Part 3 – Floor Waxes (contd)

Type of Floor	Floor Waxes to Use	Floor Waxes to Avoid
Glass	None	Solvent-based waxes and water emulsion floor waxes

Plastic seamless:
  Screeded
  Self-levelling
  Decorative
	Generally none, but water emulsion floor waxes can be used, particularly on decorative floors, if required	Solvent-based waxes

Anti-static:
  Terrazzo
  Magnesite
  Linoleum
  Rubber
  Flexible PVC
	None	Solvent-based waxes and water emulsion floor waxes

APPENDIX IV

GLOSSARY OF TECHNICAL TERMS*

Abrasive Nylon Mesh Discs These discs are circular and are supplied in a wide range of diameter to fit most makes of floor maintenance machine. They consist of nylon fabric mesh coated with resin-bonded silicon carbide.

Accelerator A substance which increases the speed of a chemical reaction. By common use, the name has become associated with two-pot surface coating materials. The accelerator usually occupies the smaller container and must be added to the larger container, the base, before use.

Acid See *pH*

Acrylic Resins Manufactured from acrylic acid. They are transparent, water-white and thermoplastic. An acrylic resin conveys the characteristics of toughness, lightness of colour and excellent water resistance.

Alkali See *pH*

Alakli-soluble Resins A resin soluble in an alkaline solution. They are widely used in water emulsion floor waxes.

Alkyd Varnishes Manufactured from glycerol. They are normally pale in colour and dry rapidly to a glossy, durable film with excellent adhesion. Alkyd varnishes are used in many interior, exterior and stoving paints, and to a lesser extent in floor seals.

All-resin Emulsion Wax An emulsion wax manufactured entirely from resin constituents. The term applies particularly to those emulsion waxes consisting of a synthetic wax, which could be called a resin, an alkali-soluble resin and a polymer, which is also a resin. The term all-resin is intended to distinguish this type of material from an emulsion floor wax containing a natural wax, an alkali-soluble resin and a polymer resin.

Base One component of a two-pot surface coating material. It usually occupies the larger container of the two-pot material. It will not, by itself, form a film and requires the addition of an accelerator before use.

*From *Maintenance of Floors and Floor Coverings,* 7th edn, Russell Kirby Ltd (1965)

APPENDIX IV

Buffable Floor Wax A term given to a floor wax which, when buffed, will give a greater gloss than when not. The waxes can be rebuffed from time to time, as required.

Button Polish A solution of a button lac in an alcohol solvent, usually of the methylated spirit type. Button lac is produced from shellac, the excretion of an insect and is so called because, when the raw shellac is refined, the end product has the appearance of buttons.

Catalyst A chemical substance used to accelerate a chemical reaction without itself being permanently changed.

Caustic Destructive or corrosive to living tissue. This term is usually met in connection with caustic soda and caustic potash, two very strong alkaline materials.

Compatibility The tolerance of one dissolved substance towards another. For example, in the process of applying a new seal on to an old one a greater degree of intercoat adhesion is achieved if the two seals are compatible. If the new seal is incompatible with the old seal, special preparation may be necessary to ensure that the new seal will adhere satisfactorily to the old.

Copolymer A very large complex molecule formed by the reaction together of a great number of small molecules of different types. An example is vinyl acetate-acrylate copolymer, a material used in adhesives.

Cure Frequently used in the same sense as the word harden. For example, a film that has fully hardened can be said to be fully cured. The word is generally used in connection with materials hardened by artificial means such as a chemical reaction or stoving at high temperature. It is not used in connection with air-drying materials, for example, oleo-resinous seals.

Detergent A cleansing agent, which may be solvent- or water-based, for removing dirt, etc. It has the advantage over soap because it is just as effective in hard water as in soft, and does not form scums.

Dilute The verb to dilute means to reduce in strength by addition of water or other appropriate solvent

Disinfectant Any compound which will destroy microorganisms. Carbolic acid (phenol) is one of the best known. Recent developments in this field have produced a large number of stronger disinfectants which are both more effective and safer to handle.

APPENDIX IV

Dry-bright Normally refers to a water-based floor wax which, on application, will dry with a glossy appearance. Dry-bright floor waxes are also known as self-gloss emulsion waxes.

Driers Used to accelerate the drying, or hardening process, particularly in air-drying seals.

Drying The process of hardening. Two stages are normally apparent in the drying process:

(a) *Touch dry* The stage at which the film will not mark when pressed lightly with a finger. At this stage the surface has hardened to the extent that it will not retain any dust or dirt that might settle upon it.

(b) *Hard dry* The stage at which the seal or dressing is sufficiently hard to withstand traffic.

Dusting This term is normally applied to concrete floors, and refers to the disintegration of the surface layer of concrete into very fine particles of dust. Almost all concrete floors dust to a greater or lesser extent, depending upon the concrete mix and type and volume of traffic.

Eggshell Finish Subdued gloss of a surface coating material.

Emulsifying Agent A chemical used in the preparation of emulsions to prevent the components from separating. It is used normally only in small quantities.

Emulsion A very fine suspension of one liquid in another with which it is not miscible. Oil and water are not normally miscible and will separate if blended together. They can, however, be emulsified by the use of emulsifying agents which suspend one liquid in another. By common use the word has also come to mean the suspension of a wide range of solid materials in water. For example, although wax is a solid, a suspension of wax in water is called water/wax emulsion.

Emulsion Waxes

(a) *Two component systems* A blend of water/wax emulsion and an alkali-soluble resin, or shellac. They may or may not dry with a glossy appearance. An increased gloss can be obtained by buffing.

(b) *Three component systems* A blend of water/wax emulsion, an alkali-soluble resin or shellac and a synthetic polymer resin emulsion. Examples of polymer resins commonly used in the polish industry are polystyrene and acrylates. The water/wax emulsion, alkali-soluble resin and synthetic polymer resin emulsion can be blended in almost any proportions to give emulsion waxes with a wide variety of properties.

APPENDIX IV

Epoxy Resin A synthetic resin manufactured essentially from petroleum derivatives. It is usually supplied in a two-pot form when used in a floor seal. The base component consists of the epoxy resin; the accelerator may be one of a variety of chemicals. In a solvent-free form it is also used for floor laying purposes.

Etching The process of forming small cavities in a surface by the use of a chemical reagent. For example, when sealing concrete floors it is often desirable to etch the surface with an acid. The cavities so formed enable the seal to penetrate further thus ensuring a greater degree of adhesion.

Film A very thin layer of a substance which, in a floor seal, is usually between 0·13 mm ($\frac{5}{1000}$ in) and 0·25 mm ($\frac{10}{1000}$ in) thick.

Finishing Coat This term is normally applied to a surface coating material used as the top coat of a painting or sealing system usually over a priming coat or undercoat.

Flashing A phenomenon associated with matt paints and seals. It describes the alternate matt and gloss striation effects sometimes left by brushmarks, instead of the uniform matt finish which should be obtained.

Flash Point The temperature at which vapour from a liquid will ignite when exposed to a small flame or spark. The lower the temperature at which ignition takes place, the more flammable is the liquid. For example, acetone, which has a flash point of $-17·8°C$ ($0°F$) will ignite below ordinary room temperature 18·3°C (65°F) and is, therefore, very highly flammable; white spirit, on the other hand, has a flash point of 41·1°C (106°F), and therefore requires the temperature to be raised before it will ignite.

Freeze-thaw Stability This property is normally associated with water emulsion floor waxes and water paints and is the resistance of the material to repeated freezing and thawing. One complete freeze-thaw cycle consists of lowering the temperature of the material until it freezes, holding it at that temperature for a specified period and then allowing it to warm to room temperature, when the material again becomes liquid. This can be repeated as required. When a material fails a freeze-thaw stability test, solid ingredients in the emulsion separate from the liquid forming a hard mass. The material is then in an unusable condition. Depending upon the type of emulsion, a material may be completely freeze-thaw stable over repeated cycles, stable over a limited number of cycles or completely unstable when frozen and thawed once.

Friction The resistance to motion which is called into play when it is attempted to slide one surface over another.

APPENDIX IV

Germicide See *Disinfectant.*

Gloss A shiny surface given by surface coating materials.

Hardener See *Accelerator.*

Intercoat Adhesion The bonding together of two coats, one upon the other, of surface coating materials.

Lacquer The correct definition of a lacquer is 'a solution of film-forming substances in volatile solvents'. Drying takes place entirely by evaporation of solvent, leaving the original film-forming substances as a thin film on the surface.

Levelling Also known as flow. It is the property of a surface coating material to flow out and spread itself evenly over the surface, so eliminating applicator or brush marks.

Liquid Wax A combination of wax and solvent, liquid at room temperature.

Matt A smooth, but dull, surface.

Metal Fibre Floor Pads These pads are circular and are supplied in a wide range of sizes to fit most makes of floor maintenance machine. They are generally manufactured in three grades – coarse, medium and fine.

Miscible Two or more liquids are said to be miscible, if, when brought together, they completely intermix to form one liquid. Two or more liquids are said to be immiscible if, when brought together, they will not intermix but separate into two or more layers.

Nylon Web Pads The pads are circular and come in many different sizes to fit most makes of floor maintenance machine. They are generally manufactured in three grades – coarse, for wax stripping, medium, for scrubbing, and fine for buffing.

Oleo-resinous A blend of oil with a resin. The oleo-resinous type is one of the oldest established seals and consists of an oil processed with a resin and combined with solvent and driers. It dries by the action of oxygen in the atmosphere causing the oil and resin to harden. This process is accelerated by the use of driers.

One-pot (*One-pack* or *One-can*) Refers to material packed in a single container and in a ready-for-use condition, without any further modification.

APPENDIX IV

Penetrating Seal A seal which will penetrate into the surface on which it is applied. Oleo-resinous seals are penetrating seals, in contrast to some plastic seals which are surface seals and do not penetrate to any great extent.

pH A method of expressing acidity and alkalinity in numerical terms. The pH scale ranges from 0 to 14. 7 is neutral and is the pH of pure distilled water. Materials with a pH below 7 are acidic, the acidity increasing as the pH decreases; materials with a pH above 7 are alkaline, alkalinity increasing as the pH increases. For example: vinegar, a weak acid, has a pH of approximately 3; hydrochloric acid, a strong acid, has a pH of between 0 and 1; ammonia, a weak alkali, has a pH of approximately 10 to 11; caustic soda, a strong alkali, has a pH of approximately 14. Acids, in general, are harmful to flooring surfaces. Alkalis will not harm floors when used correctly, but floors treated with strong alkalis must always be well rinsed to ensure that all traces are removed.

Phenolic Resin A synthetic resin manufactured basically from phenol. Widely used in many surface coating materials, for example oleo-resinous seals.

Pigment A solid colouring matter which forms a paint when mixed with a suitable liquid. Pigment not only gives the paint its colour, but also its opacity or hiding power.

Plastic A material which will soften when heated. Plastic materials can be thermoplastic or thermosetting. Thermoplastic materials can be heated and cooled repeatedly without detrimental effect. Thermosetting materials are compositions which undergo chemical change when heated and cannot be reheated without causing damage.

Plastic Seals
    (*a*) *One-pot* The description is commonly given to those plastic seals which do not contain a drying oil, and dry by evaporation of solvent or by a chemical reaction which is activated by evaporation of solvent.
    (*b*) *Two-pot* This description is commonly given to those plastic seals which require the blending together of two components prior to use.

Polymer A very large, complex molecule formed by the reaction together of a great number of small molecules of the same type. Examples are polystyrene and polyacrylate, materials often used in water-based waxes.

APPENDIX IV

Polystyrene Resin A compound formed by the polymerisation of a resin, styrene. In emulsion waxes polystyrene imparts excellent gloss, hardness and levelling.

Polyurethane A polymer formed as the result of a chemical reaction between two types of compound, namely an isocyanate and a form of polyester. Among many other applications, polyurethanes are used in floor seals and paints. For these purposes they are normally supplied in three different forms:

(*a*) *Two-pot* The base component is the polyester and the accelerator, or hardener, the isocyanate. The latter is extremely sensitive to water and moisture vapour in the atmosphere and must be protected from them during storage. Once the base and accelerator are mixed, a chemical reaction is started which stops only when the material has solidified.

(*b*) *One-pot, Oil-modified* In these materials the urethane has already been produced and is further combined with an oil or varnish. They are often referred to as urethane oils. Drying takes place by oxidation of the oil or varnish component.

(*c*) *One-pot, Moisture-cured* These materials consist of urethane, already produced, but with an excess of isocyanate present. Once applied, the excess isocyanate attracts water vapour from the atmosphere and hardens the material. The rate of drying will, therefore, depend largely on the humidity, but in this climate there is sufficient moisture in the atmosphere to effect a complete hardening of the film.

Pot-life This term refers to two-pot materials and is the period during which the material is usable once the base and accelerator components have been blended together. After this period, the material will have thickened to such an extent that it cannot be applied satisfactorily.

Priming Coat This is the first coat applied to previously untreated surfaces. It provides a foundation on which the durability of the finished system largely depends. For example, on wood surfaces the primer is required to be absorbed into the surface in order to obtain a 'key' for subsequent coats. On cement, plaster and concrete surfaces the primer is formulated so that it will resist chemical attack by the alkaline ingredients of the surface to which it is applied.

PVC (polyvinyl chloride) Floor Coverings There are two main types of PVC (polyvinyl chloride) floor coverings in use today. These are, first, flexible PVC flooring and second, PVC (vinyl) asbestos floor tiles. Flexible PVC flooring is supplied in sheet or tile form and both have a smooth wearing surface. PVC (vinyl) asbestos is usually supplied in tile form. They are similar in composition, the main difference being that

APPENDIX IV

PVC (vinyl) asbestos contains asbestos fibre, which is not present in flexible PVC floor coverings. Asbestos fibre is adversely affected by many solvents, for example white spirit. While the above are the two main types of PVC floor covering in use to-day, tiles with intermediate characteristics are produced by varying the relative amounts of polyvinyl chloride and asbestos fibre. It is, therefore, frequently very difficult to distinguish one type of tile from another.

Rafting This is a phenomenon which can occur with sealed new wood block floors, if they have not been properly prepared. It is the movement of a large number of blocks simultaneously, causing a crack to appear in the floor. This can be caused if the blocks are subjected to considerable change in moisture content, causing them to shrink excessively, while at the same time they are tightly bonded together by the seal. Instead of swelling and shrinking individually into existing gaps, the blocks move as a mass thereby causing the crack to appear.

Reaction Coating A surface coating formed by reaction between two or more chemicals. For example, the film formed by a polyurethane two-pot seal is a reaction coating, because a chemical reaction takes place when the base and accelerator are mixed together.

Resin A resin can be naturally occurring or synthetic and is characterised by being insoluble in water but soluble in a wide range of solvents, for example, white spirit. Naturally-occurring resins are adhesive substances obtained from sources such as pine trees. Synthetic resins are made by chemical means. There are many different resins in use in the industry, for example, phenolic and polystyrene resins.

Rivelling This phenomenon can best be described as severe wrinkling. It normally takes place where seal has been applied too thickly and where the surface has dried quicker than the body of the seal, causing the surface to wrinkle.

Seal A floor seal can be described as a semi-permanent finish which, when applied to a floor, will prevent the entry of dirt and stains, liquids and foreign matter.

Self-gloss See *Dry-bright*.

Shelf-life The period during which a finished product is in a usable condition in its container. After this period the material may be unsuitable for use due to a variety of reasons, for example, thickening in the tin, excessive rusting of the tin, decomposition due to bacterial attack, etc.

APPENDIX IV

Skin A thick layer of material over the surface coating material, for example a paint or floor seal, formed by the oxidation of the surface layer.

Softwood Softwood is wood which belongs to the order Coniferae, or conifers, which includes for example spruce, Douglas fir and longleaf pitch pine.

Solid Content The total solid constituents, usually expressed as a percentage, remaining when all solvents are removed from a material.

Solvent Any liquid which will dissolve a solid is a solvent for that solid. Although water is a solvent for many materials, by common use the word solvent has come to mean liquids other than water. White spirit, for example, is a solvent for many resins. Solvent is normally included in a seal to aid application by enabling the material to be spread more easily.

Specific Gravity is the ratio of the mass of a given volume of a substance to the mass of an equal volume of water at a temperature of 4°C. For example, white spirit has a specific gravity of 0·787, compared with that of 1·000 for water. White spirit is, therefore, lighter in weight than the same volume of water. Specific gravity is measured in g/ml. To convert to lb/gal, multiply the specific gravity by ten.

Synthetic Artificial or man-made. Not derived immediately from naturally-occurring materials.

Thermoplastic See *Plastic*

Thinner A liquid added to a paint or varnish to facilitate application. For example, xylene is a thinner widely used in polyurethane seals. Once the seal is applied, the thinner evaporates.

Toxic Poisonous. Toxicity is the degree to which a substance is poisonous.

Translucent A material which is translucent will allow light to pass through it, without being transparent.

Two-pot (*Two-pack* or *Two-can*) Refers to materials supplied in two separate containers. The contents of one container must be added to the other and the blended material thoroughly mixed before use. The larger container generally holds the base and the smaller the accelerator or hardener.

APPENDIX IV

Urea-formaldehyde A synthetic resin manufactured by heating together two chemicals, urea and formaldehyde. Urea-formaldehyde is widely used in both one- and two-pot seals. The seals cure by the action of an acid catalyst, which, in a two-pot seal is the accelerator, or hardener, component. They have the characteristic of being almost water-white in colour.

Vinyl Resin A synthetic resin used in the manufacture of many water emulsion paints, floor coverings, etc.

Vinyl Floor Covering See *PVC (polyvinyl chloride) Floor Coverings.*

Viscosity This is the resistance of a liquid to flow; the greater the resistance, the higher the viscosity. For example, a thick engine oil has a greater viscosity than has a thin cycle oil. Viscosity rapidly decreases with increase in temperature.

Wax
    (a) *Natural* A solid material, chemically related to fats. There is a very wide range of naturally-occurring wax. Examples are beeswax, a soft wax, produced from the sugar of food eaten by bees, formed in the bee's stomach, and carnauba wax, a hard wax, produced from the leaves of trees found mainly in Brazil.
    (b) *Synthetic* There is also a very wide range of synthetic wax. A well known example of a soft wax is paraffin wax, derived from petroleum. Polyethylene is an example of a harder synthetic wax frequently used in both water- and solvent-based polishes.

Wetting Agent This is used to reduce the surface tension between a solid and a liquid. In detergents, a wetting agent is included to loosen dirt from the surface to which it is attached.

White Spirit A solvent derived from the distillation of petroleum and generally known as turpentine substitute. It is widely used in the polish, paint and varnish industries.

APPENDIX V

CONVERSION TABLES

MEASURES OF WEIGHT – AVOIRDUPOIS AND COMMERCIAL

1 ounce (oz)   = 16 drams (dr)
1 pound (lb)   = 16 oz
1 stone (st)   = 14 lb
1 quarter (qr)   = 2 st = 28 lb
1 hundredweight (cwt) = 4 qr = 112 lb
1 ton   = 20 cwt = 2 240 lb

METRIC UNITS OF WEIGHT

1 kilogramme (kg) = 1 000 grammes (g)
1 quintal (q)   = 100 kg
1 tonne   = 1 000 kg

EQUIVALENT UNITS OF WEIGHT

Avoirdupois   *Metric*

1 ounce   = 28·35 g
1 pound   = 453·6 g = 0·454 kg
1 stone   = 6·35 kg
1 quarter   = 12·7 kg
1 hundredweight   = 50·802 kg
1 ton (2 240 lb) = 1 016·04 kg = 1·016 tonnes

Metric   *Avoirdupois*

1 gramme   = 0·035 oz
1 kilogramme = 2·2046 lb = 35·274 oz
1 quintal   = 1·968 cwt = 220·462 lb
1 tonne   = 0·984 tons

APPENDIX V
U.S.A. UNITS

The fundamental units are the metre and the gramme, but the pound and the yard are also in general use. A U.S. ton, or short ton, of 2 000 lb is widely used.

U.S.A.	British	Metric
1 short ton	= 0·8928 tons = 2 000 lb	= 907·184 kg
1 short cwt	= 0·8928 cwt = 100 lb	= 45·359 kg
1·12 short tons	= 1 ton = 2 240 lb	= 1 016 kg = 1·016 tonnes
1·12 short cwt	= 1 cwt = 112 lb	= 50·802 kg
1·1023 short tons	= 0·9842 tons = 19·69 cwt	= 1 tonne = 1 000 kg

LENGTH

1 statute mile = 1 760 yards (yd) = 5 280 feet (ft) = 63 360 inches (in)
1 yard = 0·9144 metres (m) = 91·44 centimetres (cm) = 914·4 millimetres (mm)

1 kilometre (km) = 1 000 (m) = 100 000 (cm)
1 metre = 1·0936 yd = 39·37 in = 3·2808 ft

AREA

1 square yard (yd^2) = 9 ft^2 = 0·836 m^2
1 square foot (ft^2) = 144 in^2 = 929 cm^2
1 square metre (m^2) = 1·196 yd^2 = 10·764 ft^2

CAPACITY

1 cubic yard (yd^3) = 27 ft^3 = 0·7646 m^3
1 cubic metre (m^3) = 35·315 ft^3 = 1·30795 yd^3
1 Imperial gallon (Imp. gal) = 4 quarts = 8 pints = 4·546 litres
1 litre (l) = 1·76 pints = 0·22 Imp. gal = 1 000 millilitres (ml)
1 pint = 0·568 l = 0·125 Imp. gal

U.S.A. Gallons

1 U.S. gal = 0·8327 Imp. gal = 3·785 l
1 Imp. gal = 1·2 U.S. gal = 4·546 l
1 litre = 0·264 U.S. gal = 0·22 Imp. gal

APPENDIX V
THERMOMETRICAL

Fahrenheit: Freezing Point = 32°
Boiling Point = 212°

Centigrade: Freezing Point = 0°
Boiling Point = 100°

To convert Fahrenheit to Centigrade

$$(°F - 32) \times 5/9 = °C$$

e.g. To convert 68°F to Centigrade

$$(68 - 32) \times 5/9 = \frac{36 \times 5}{9} = 20°C$$

To convert Centigrade to Fahrenheit

$$(°C \times 9/5) + 32 = °F$$

e.g. To convert 20°C to Fahrenheit

$$\left(\frac{20 \times 9}{5}\right) + 32 = 36 + 32 = 68°F$$

OTHER USEFUL FACTORS

¼ Imp. gal = 40 fluid oz
1 Imp. pint = 20 fluid oz
1 tablespoonful = ½ fluid oz
1 teaspoonful = ⅛ fluid oz
1 litre = 35·2 fluid oz
1 fluid oz = 28·4 ml
1 gal water (at 62°F) weighs 10 lb
1 litre water (at 62°F) weighs 2·2 lb

APPENDIX V
METRIC CONVERSION TABLE

ft	in	mm*	ft	in	mm*
–	1/16	2	–	1	25
–	1/8	3	–	2	51
–	3/16	5	–	3	76
–	1/4	6	–	4	102
–	5/16	8	–	5	127
–	3/8	10	–	6	152
–	7/16	11	–	7	178
–	1/2	13	–	8	203
–	9/16	14	–	9	229
–	5/8	16	–	10	254
–	11/16	17	–	11	279
–	3/4	19	1	0	305
–	13/16	21	2	0	610
–	7/8	22	3	0	914
–	15/16	24			

* Correct to the nearest whole millimetre.

INDEX

Abrasive nylon mesh discs, use on
  cork, 31, 32
  magnesite, 36
  wood, 24
  wood composition, 28
Acid, effect on
  asphalt, 64
  concrete, 42
  cork, 33
  flexible PVC, 85
  granolithic, 42
  magnesite, 35, 37
  marble, 50
  natural stone, 54
  PVC (vinyl) asbestos, 85
  quarry tile, 57
  rubber, 90
Adamantine tiles, 55
Afromosia, 18
Alkaline detergents, 2
  use on,
    asphalt, 67
    clay tiles, 59
    concrete, 44
    cork, 33
    flexible PVC, 88
    granolithic, 44
    linoleum, 74
    magnesite, 37
    natural stone, 54
    plastic seamless, 122
    PVC (vinyl) asbestos, 88
    rubber, 93
    terrazzo, 48
    thermoplastic tiles, 82
    wood, 26
Aluminium floors, 115
Anti-static agents, 102
Anti-static floors, 123
  characteristics, 125
  flexible PVC, 124
  linoleum, 124
  magnesite, 124

Anti-static floors—*contd.*
  maintenance, 125
  rubber, 124
  terrazzo, 124
Ashburton limestone, 51
Ashburton marble, 49
Asphalt, 62
  characteristics, 63
  detergents, use on, 67
  maintenance, 64
  sealing, 65
  tiles, 77
  waxing, 66
Asphaltic cement, 63
Axminster carpets, 97, 98

Beech, 18
Belgian black marble, 49
Bitumen, 63
Blue rag limestone, 51
Brick, 59
  maintenance, 60
  sealing, 60
  waxing, 60
Brussels carpet, 98
Button polish, 31

Carpet, 94
  anti-static agents, 102
  Axminster, 97, 98
  backing, 97
  Brussels, 98
  characteristics, 100
  classification, 98
  deep steam cleaning, 107
  discolouring, 104
  dry-foam shampoo, 110
  dry-powder cleaner, 108
  dye, 98
  electrostatically flocked, 98, 99
  fibres, natural, 95, 96

INDEX

Carpet–*contd.*
 fibres, synthetic, 96
 fluffing, 103
 hair tile, 104
 knitted, 98, 99
 liquid shampoo, 108
 maintenance, 104
 mechanical equipment, use of, on, 106
 moisture content, 96
 needleloom, 98, 99
 non-static, 103
 pile-bonded, 98, 99
 rotting, 100
 shading, 103
 shooting, 103
 shrinkage, 104
 silicone treatment, 111
 soil, types of, 105
 soil removal, 105
 spot removers, 110
 sprouting, 103
 stain removal, 111
 static electricity, 101
 sweepers, 106
 tapestry, 98
 tiles, 100
 tufted, 98, 99
 vacuum cleaners, use of, on, 106
 Wilton, 97, 98
Caustic detergents, 2
 effect on wood, 26
Cement latex, 60
 sealing, 61
 waxing, 61
Ceramic tiles, 54
Characteristics of
 anti-static floors, 125
 asphalt, 63
 carpet, 100
 concrete, 40
 cork, 29
 cork carpet, 75
 flexible PVC, 84
 granolithic, 40
 iron floors, 114
 linoleum, 69
 magnesite, 34
 marble, 49
 natural stone, 51
 PVC (vinyl) asbestos, 84
 quarry tiles, 56
 rubber, 89
 steel floors, 114

Characteristics of–*contd.*
 terrazzo, 46
 thermoplastic tiles, 78
 wood, 17
Chinese carpets, 95
Clay tiles, 54
 detergents, use on, 59
 sealing, 58
 waxing, 58
Concrete, 38
 characteristics, 40
 detergents, use on, 44
 dusting, 40
 etching, 42
 maintenance, 41
 sealing, 41
 waxing, 44
Conductive floors, 123
Cork, 28
 acid, effects of, 33
 characteristics, 29
 maintenance, 30
 sealing, 31
 waxing, 32
Cork carpet, 74
 characteristics, 75
 detergents, use on, 77
 maintenance, 75
 sealing, 76
 waxing, 76

Detergents,
 alkaline, 2
 caustic, 2
 crystals, 3, 44
 neutral, 2
 rinsing, 3
 solvent wax removers, 3
 types, 1
Dirt, definition of, 1
 grease, 1
 nature of, 1
 organic, 2
 particulate, 1
Douglas fir, 22
Dry foam carpet shampoo, 110
Dry residue carpet shampoo, 109
Dust allaying oil, 21
Dusting, concrete and granolithic, 40

Electrostatically flocked carpet, 98, 99
Encaustic tiles, 55

INDEX

End-grain paving, 14, 17
Enstone limestone, 51
Epoxy seamless floors, 117
Etching of concrete, 42

Faience tiles, 55
Flexible PVC, 82
   characteristics, 84
   detergents, use on, 88
   maintenance, 85
   sealing 86
   waxing, 86
Floor seals, 4
   definitions of, 4
   surface preparation, 8
   types of, 4
Floor waxes, 8
Flux oil, 63
Foam cleaning, 12

Gaboon, 18
Glass floors, 116
   maintenance, 116
Glass mosaics, 116
Glazed tiles, 55
Granite, 50
Granolithic, 38
   characteristics, 40
   detergents, use on, 44
   dusting, 40
   etching, 42
   maintenance, 41
   sealing, 41
   waxing, 44
Gurjun, 22
Gymnasium oil, 21

Horton rag limestone, 51
Hydrochloric acid, 40, 42
   use on,
      natural stone, 52
      quarry tiles, 57, 58

Industrial pavoirs, 55
Iroko, 18
Iron floors, 113
   characteristics, 114
   maintenance, 114

Keruing, 22
Knitted carpet, 98, 99

Lake asphalt, 62
Limestone, 50
Linoleum, 68
   characteristics, 69
   detergents, use on, 74
   maintenance, 70
   sealing, 71
   waxing, 72
Liquid wax, 8

Magnesite, 33
   acid, effects of, 35, 37
   characteristics, 34
   detergents, use of, 37
   maintenance, 35
   sealing, 36
   waxing, 36
Maintenance of
   aluminium floors, 116
   anti-static floors, 125
   asphalt, 64
   brick, 60
   carpet, 104
   concrete, 41
   cork, 30
   cork carpet, 75
   flexible PVC, 85
   glass floors, 116
   granolithic, 41
   iron floors, 114
   linoleum, 70
   magnesite, 35
   marble, 50
   natural stone, 52
   plastic seamless, 120
   PVC (vinyl) asbestos, 85
   quarry tiles, 57
   rubber, 90
   steel floors, 114
   terrazzo, 47
   thermoplastic tiles, 79
   wood, 20
   wood composition, 28
Maple, 18
Marble, 49
   characteristics, 49
   maintenance, 50
Marble linoleum, 69
Mastic asphalt, 62

INDEX

Moiré linoleum, 69
Mosaic tiles, 54

Natural carpet fibres, 95
Natural rubber, 88
Natural stone, 50
  characteristics, 51
  detergents, use on, 53
  maintenance, 52
  sealing, 53
  waxing, 53
Needleloom carpet, 98, 99
Neutral detergents, 2
  use on,
    asphalt, 67
    clay tiles, 59
    concrete, 44
    cork, 32
    cork carpet, 77
    flexible PVC, 88
    granolithic, 44
    linoleum, 74
    magnesite, 37
    natural stone, 54
    plastic seamless, 122
    PVC (vinyl) asbestos, 88
    terrazzo, 48
    thermoplastic tiles, 82
    wood, 26
Neutralising solution, 3
Non-static carpets, 103

Oak, 18
Oleo-resinous seals, 4
One-pot seals, 5

Parquet, 14, 16
Paste wax, 8
Paver tiles, 55
Pavoirs, industrial, 55
Persian carpets, 95
pH scale, 2
Pigmented seals, 7
Pile-bonded carpets, 98, 99
Pitch mastic, 62
Plastic seals, 1-pot, 5
Plastic seals, 2-pot, 5
Plastic seamless floors, 117
  detergents, use on, 122
  maintenance, 120
  sealing, 121
  waxing, 122

Polyester seamless floors, 117
Polyurethane seals, 1-pot, 5
Polyurethane seals, 2-pot, 6
  pigmented, 7
Polyurethane seamless floors, 117
Poultice, use on,
  natural stone, 53
  quarry tiles, 57
  terrazzo, 47
PVC (vinyl) asbestos, 82
  characteristics, 84
  detergents, use on, 88
  maintenance, 85
  sealing, 86
  waxing, 86

Quarry tiles, 54
  characteristics, 56
  maintenance, 57
Quartzite, 50

Rafting, 19, 23
Ribbed rubber tiles, 89
Rock asphalt, 62
Roman stone, 49
Rubber, 88
  acids, effects of, 90
  characteristics, 89
  detergents, use on, 93
  maintenance, 90
  oxygen, effects of, 91
  sealing, 92
  sunlight, effects of, 91
  waxing, 92

Sanding, cork, 31, 32
  wood, 14, 22, 23
Sandstone, 50
Sapele, 18
Screeded plastic floors, 118
Seals, use on,
  asphalt, 65
  brick, 60
  cement latex, 61
  clay tiles, 58
  concrete, 41
  cork, 31
  cork carpet, 76
  flexible PVC, 86
  linoleum, 71
  magnesite, 36

INDEX

Seals, use on—*contd.*
  natural stone, 53
  plastic seamless, 121
  PVC (vinyl) asbestos, 86
  rubber, 92
  terrazzo, 47
  thermoplastic tile, 80
  wood, 22
  wood composition, 27
Seamless plastic floors, 117
Secret nailing, 15
Self-levelling plastic floors, 118
Shampoo for carpets, 108
Sicilian marble, 49
Silicate dressings, 7
  use on,
    concrete, 41
    granolithic, 41
Silicone treatment for carpets, 111
Slate, 50
Solvent based detergent wax
  removers, 3, 45
Solvent based wax, 8
  removal of, 11
Spot removers for carpets, 110
Spray cleaning, 12
Static electricity, 101, 123
Steel floors, 113
  characteristics, 114
  maintenance, 114
Sycamore, 18
Synthetic rubber, 88
Synthetic carpet fibres, 96, 97

Tapestry carpet, 98
Teak, 18, 22
Terrazzo, 45
  characteristics, 46
  detergents, use on, 48
  maintenance, 47
  sealing, 47
  waxing, 48
Tessellated tiles, 55
Thermoplastic tiles, 77
  characteristics, 78
  detergents, use on, 82
  maintenance, 79
  sealing, 80
  waxing, 81
Tiles,
  adamantine, 55
  ceramic, 54
  clay, 54

Tiles—*contd.*
  encaustic, 55
  faience, 55
  glazed, 55
  mosaic, 54
  paver, 55
  quarry, 54
  tessellated, 55
  vitreous, 54
Travertine, 49
Trinidad lake asphalt, 62
Tufted carpets, 98, 99
Turk's head brush, 43
Turkish carpets, 95
Two-pot plastic seals, 5

Vinegar, 3
Vitreous tiles, 54
Vacuum cleaner, 106

Water based seals, 7
Water based wax, 9
  dry-bright, 10
  fully buffable, 10
  removal of, 11
  semi-buffable, 10
  types of, 10
Wax, liquid, 8
Wax, paste, 8
Wax, water based, 9
Waxes, use on,
  asphalt, 66
  brick, 60
  cement latex, 61
  clay tiles, 58
  concrete, 44
  cork, 32
  cork carpet, 76
  flexible PVC, 86
  granolithic, 44
  linoleum, 72
  magnesite, 36
  natural stone, 53
  plastic seamless, 122
  PVC (vinyl) asbestos, 86
  rubber, 92
  terrazzo, 48
  thermoplastic tiles, 81
  wood, 25
  wood composition, 28
Western hemlock, 18

INDEX

Wilton carpets, 97, 98
Wood, 13
    acid, effects of, 19
    alkali, effects of, 19
    blocks, 15
    characteristics, 17
    classes of, 14
    colour, 18, 25
    detergents, use on, 26
    types of, 14
    hardwood, 14
    heartwood, 14
    maintenance, 20
    moisture content, 19

Wood–*contd.*
    mosaic, 17
    parquet, 16
    rafting of blocks, 19
    sapwood, 14
    sealing, 22
    softwood, 14
    strip and board, 15
    water, effects of, 19
    waxing, 25
Wood composition, 27
    maintenance, 28
    sealing, 28
    waxing, 28